Hans Christian von Baeyer is Chancellor Professor of Physics
at the College of William and Mary, Williamsburg, Virginia.
In the course of his career he has been elected Fellow of the
American Physical Society, served as Director of the Virginia
Associated Research Campus – an interdisciplinary research
laboratory which became the nucleus of the Jefferson Lab
(the world's premier electron accelerator facility for nuclear
physics) – and garnered numerous awards for his scientific
writing. His previous books include *Taming the Atom*, *Rainbows, Snowflakes and Quarks* and *Warmth Disperses and Time
Passes* (formerly *Maxwell's Demon*).

Information

The New Language of Science

Hans Christian von Baeyer

PHOENIX

A PHOENIX PAPERBACK

First published in Great Britain in 2003
by Weidenfeld & Nicolson
This paperback edition published in 2004
by Phoenix,
an imprint of Orion Books Ltd,
Orion House, 5 Upper St Martin's Lane,
London WC2H 9EA

A CIP catalogue record for this book
is available from the British Library.

ISBN 0 75381 782 9

Printed and bound in Great Britain by
Clays Ltd, St Ives plc

www.orionbooks.co.uk

To the millennial generation: Irene, Teo, and Anna

Contents

Quantum Information

Work in Progress

Prologue

Really Big Questions

For his birthday, John Archibald Wheeler got an argument. The dean of American theoretical physicists turned ninety on 9 July 2001, but the international symposium called to mark the event did not take place until the following spring. In the intervening eight months the invited guests, plus anyone else who dared to join in, engaged in a freewheeling, knock-down, drag-out e-mail debate on issues of interest to the great man, ranging from string theory to the nature of time. Its full text, as rich in complexity and passion as it was short on conclusions, was printed out and bound as a gift, filling 236 pages and delighting its recipient.

The celebratory symposium itself was sponsored by a consortium of organizations, including the John Templeton Foundation for promoting dialogue between science and religion, and furnished with an imposing title worthy of the grandeur of Wheeler's vision: *Science and Ultimate Reality*. Three hundred of his colleagues, students, and friends, together with a sprinkling of Nobel laureates and guests from as far away as Africa and Australia, many of whom had previously met only in cyberspace, gathered at a posh conference centre in his home town of Princeton, New Jersey, for three days of scientific lectures and complimentary speeches organized around an elegant banquet. President George W. Bush and Pope John Paul II sent congratulations.

Wheeler is a bit hard of hearing now, his handwriting is shaky, and his gait has lost its spring, but throughout the meeting he sat in the centre of the front row with a youthful, impish smile crinkling the corners of his eyes. When he stood up to make a remark from time to time, he demonstrated that his wit has not abandoned him; but at the end of the festivities, when he thanked the assembled guests for honouring him with their presence, and inspiring him with fresh ideas, we realized that he had it back-to-front: the honour was all ours. The inspiration Wheeler has bestowed upon scientists throughout the world in his seventy-year career is beyond measure; we owe him a debt we cannot repay.

John Wheeler is a physicist's physicist. He is not widely known outside the physics community, but in the profession his has been a household name for many decades. As a young man he worked with the two dominant figures of modern physics, Albert Einstein and Niels Bohr. Just as important as his innovative contributions to the theories pioneered by those two giants – relativity and quantum mechanics, respectively – has been his powerful influence as teacher and writer of textbooks. The most famous of the many physicists he mentored was the celebrated *enfant terrible* of American physics, the late Nobel laureate Richard Feynman.

As a teacher, Wheeler understood the magic of words. In order to seduce the physics community into paying serious attention to the collapse of stars, for example, he coined the phrase 'black hole' – a ploy that succeeded beyond all expectation. In his later years, however, as his reputation grew and his essays and lectures began to be directed at wider audiences, Wheeler started to address questions at the interface of physics and philosophy – profound problems such as the interpretation of quantum mechanics and the origin of the universe. In an effort to couch his ideas in accessible and unforgettable language, he developed a unique and somewhat oracular style of expression featuring what he calls 'Big Questions'. The most significant of these, the Really Big Questions (or RBQs), suggested the organization of topics at the birthday symposium. One by one the lecturers bore personal witness to the inspiration they had drawn from these cryptic maxims.

Wheeler's RBQs are frankly metaphysical in character, and sufficiently succinct to adorn bumper stickers, though pondering them in traffic would be ill advised. Five of them stand out:

HOW COME EXISTENCE?
WHY THE QUANTUM?
A PARTICIPATORY UNIVERSE?
WHAT MAKES MEANING?
IT FROM BIT?

The first question paraphrases the famous religious and philosophical puzzle: 'Why is there something rather than nothing?' Careers have been devoted to chasing after that one. The World Wide Web is abuzz with it.

The second question represents the search for a simple, convincing argument that the world of atoms must be controlled not by the classical physics of everyday life, but by the counterintuitive rules of quantum mechanics. With characteristic courage Wheeler eschews the scientist's customary pusillanimous *how?* (to which we know the answer), and goes straight for the more elusive *why?*

'Participatory universe' is Wheeler's catchphrase for the controversial notion that the universe is not entirely 'out there' and ready to be discovered, but shaped in part by the very questions we ask about it and the information we receive in answer to them. He illustrates this proposal by telling the story of three baseball umpires who define balls and strikes according to their world views: 'I call 'em as I see 'em,' brags the first one, evidently an empiricist. 'I call 'em the way they are,' proclaims the realist. The third explains: 'They ain't nothin' until I call 'em,' making Wheeler's point.

The fourth question, 'What makes meaning?' refers to the thorny philosophical problem of defining the concept of meaning. At the same time it recalls the frustration of engineers who have at their disposal a variety of methods for measuring the amount of information in a message, but none to deal with its meaning.

The fifth and last RBQ is the most radical. The suggestion is that the material world – the IT – is wholly or in part constructed from information – the BIT. Wheeler amplifies enigmatically: 'Every it –

every particle, every field of force, even the space-time continuum itself – derives its function, its meaning, its very existence entirely – even if in some contexts indirectly – from the apparatus-elicited answers to yes-or-no questions, binary choices, *bits*.'

As I listened to the speakers at the 'Science and Ultimate Reality' symposium echoing the RBQs, even as they discussed arcane technical points of experimental and theoretical physics and cosmology, I began to understand the source of John Wheeler's impressive influence. He has achieved much more than just putting the fun back into our science, as one participant observed gratefully. In fact he has accomplished nothing less than to put metaphysics back into physics, from which it had been banned for centuries. As a student I was taught emphatically that physics tries to find the *how* of nature, leaving to philosophy and theology the task of filling in the *why*; but most people, including scientists in the privacy of their innermost thoughts, really yearn to understand why – at least at some level. Teaching students, even those who are not preparing to become scientists, about the *how*, when all they really want to know is *why*, is a frustrating experience for all teachers. Wheeler has given us permission to bring the really big questions back into the laboratory and the classroom, and he has done so not by exhortation, but by example. His oracular pronouncements do not stand alone, but are embedded in painstaking, technical contributions to conventional research. His notable successes in answering *how* questions give him licence to ask *why*. Thus Wheeler the philosopher, deriving his authority from the reputation of Wheeler the physicist, inspires his colleagues and encourages them to follow in his footsteps.

The symposium's keynote speaker, Anton Zeilinger of the University of Vienna, provided an explicit example of Wheeler's powerful influence in a talk entitled: 'Why the Quantum? It from Bit? A Participatory Universe?: Three Far-reaching, Visionary Questions from John Archibald Wheeler and How They Inspired a Quantum Experimentalist.' Zeilinger shares Wheeler's hope that '... tomorrow we will have learned to understand and express *all* of physics in the language of information'. In the last chapter of this book I shall return to Zeilinger's thesis that information is indeed the irreducible seed from which everything else grows.

Near the end of the symposium I caught up with Wheeler in the opulent lounge of the convention centre, between a pool table (traditional symbol of classical physics) and a bubbling fountain (symbol of quantum waves). After reminding him of some of our previous conversations over the past forty years, I thanked him for a postcard he had sent me in 1960, requesting a reprint of an article I had just co-authored – my first scientific paper. Even though I was baffled at the time by his interest in the subject (a fine point in the theory of how heat flows through certain obscure materials), I remember my elation at getting mail from Mount Olympus. In the lounge, Wheeler asked me about the whereabouts of my co-author and mentor Joe Callaway, and I had to bring him the sad news of Joe's death several years ago. After graduating from the College of William and Mary (now my own academic home), Joe had gone to study at Princeton and made a name for himself by finding a fatal flaw in one of Einstein's speculative attempts to unify gravity and electromagnetism. Wheeler reminisced: 'I recall Einstein's study. It must have been in the mid-fifties, shortly before his death. We were sitting together, Einstein and Joe and I, just the three of us, and we were debating the unified theory. What did he die of? Heart?' I didn't know.

Later I looked up Callaway's relativity article. It concludes with the words: 'I wish to thank Professor J. A. Wheeler and Professor A. Einstein for valuable discussions. This note was stimulated by the conflict in their views.' After four decades I finally understood Wheeler's interest in my first paper: it wasn't the subject matter that had attracted his attention, but the name of the senior author; but I wasn't disappointed. On the contrary, I felt that my bond to the community of physicists had been strengthened. From Einstein to Wheeler to Callaway to me – the handing-down of wisdom from the master to the journeyman to the apprentice in time-honoured tradition – links in the charismatic chain that ties each of us to our great predecessors.

Thus empowered, I ask two additional Really Big Questions. First, the fundamental WHAT IS INFORMATION? And then an update of Wheeler's boldest proposal, translating IT FROM BIT? into the new language of quantum information, so it becomes IT FROM QUBIT?

Background

1 Electric Rain

Information in our lives

Information gently but relentlessly drizzles down on us in an invisible, impalpable electric rain. Encoded in radio waves that fill the atmosphere, its mists fill the air, passing through the walls of our houses and penetrating our very bodies. CNN News, Australian talk shows, German soap operas, Mexican movies, and lots and lots of ads – whether quality programming or throwaway trash, they are all out there, some signals crystal clear, others hidden under impenetrable gobs of noise. To find them, all we have to do is hold out a little antenna, amplify the minute electrical impulses it catches, and convert them into sound and light.

That antenna would actually collect far more than just radio and TV shows: a babble of conversations among pilots, seamen and radio amateurs, a torrent of mobile-phone traffic, and a fog of secret military communications compete with transmissions from scientific satellites and space probes behind the clatter of countless remote-control boxes for opening garages and switching TV channels. The din is overpowering.

Then there are the wires of copper and glass. To open these pipelines, just plug in a modem and watch a flood of information from the world's uncounted electronic memories come pouring out into your laptop. Even this traffic is rapidly bursting out of the confines of cables and optical fibres, and joining the wireless uproar that surrounds us. Some is comprehensible, but most of it, such as the trillions of dollars shuffled night and day from one financial centre to another in the form of secret ciphers, makes sense only to the initiated. For better or for worse the world is awash with information.

Just as water consists of individual molecules, information comes in the form of electrical pulses represented by the numbers zero and one. Older technologies make use of more complex, analogue signals, but today information is normally encoded in the rudimentary alphabet of the computer – zeroes and ones – the fundamental stuff of information. Our familiar universe of atoms and radiation, it turns out, is suffused by an immaterial parallel universe composed of those two irreducibly primitive symbols. But a random collection of zeroes and ones is not information; it is only when those symbols are organized into distinct patterns that the information emerges, in the same way that rivers, ice cubes, clouds, and raindrops only emerge as recognizable entities when multitudes of water molecules arrange themselves in characteristic patterns. Symbols furnish the substrate – information carries the meaning.

The information universe, like the population of viruses in a body and the accumulation of compound interest, grows exponentially. Exponential growth is characterized by the remarkable property that the latest instalment is always, by itself, greater than the sum of everything that has come before. (In the exponentially growing sequence of doubling numbers 1, 2, 4, 8, 16, 32, 64 ..., notice that 4 is more than $1 + 2$, 8 is more than $1 + 2 + 4$, 16 more than $1 + 2 + 4 + 8$, and so on to infinity.) Thus, when scientists at the University of California predicted that humans and their machines will create more information in the next three years than in the preceding 300,000 years of history, they were simply describing the exponential growth of the volume of information, which doubles its value in roughly three years.

Plotted in time, the growth curve of the information universe is shaped like the trajectory of a jet fighter that takes off at a shallow slope, suddenly appears to turn a corner, and then soars with increasing steepness to unbounded heights. The initial long leg of almost imperceptible growth is usually mistaken for stagnation by all but the keenest observers. One of the people who sensed the impending eruption of information technology early on was Marshall McLuhan, who coined the maxim 'The medium is the message.' As far back as 1964 he pointed out: 'When IBM discovered that it was not in the business of making office equipment or business machines, but that it was in the business of processing information, then it began to navigate with clear vision.' Soon thereafter the explosive development of digital computing triggered the information revolution, which ushered in our computer epoch, also known as the 'digital era' or the 'information age'. Instead of ferrying people and commodities around the solar system, as we had fully expected to be doing in the twenty-first century, we traffic instead in disembodied clouds of information.

Will the exponential growth of the universe of information continue? Since real systems such as embryos and bank accounts can never grow to infinite proportions, they invariably encounter limits to growth that flatten or even reverse their plots. The curve that records a finite system's evolution has a universal shape resembling a rounded upward step – a forward-leaning S. (First mentioned by the eighteenth-century economist Thomas Malthus in reference to human populations, it is sometimes called an 'S curve'.) Its flat upper plateau, where increase yields to stability or decrease, is a consequence of the limiting mechanisms, whatever they may be.

Progress in information technology is currently only limited by engineering issues that can be overcome by clever design. It has been estimated that more fundamental limits, dictated by physical law, will not be reached until the 2020s, so for the moment exponential growth is still the rule. In the future, computer memories may well become smaller and cheaper until they reach atomic, or possibly even nuclear proportions; the range of frequencies, or channels, available for transmitting information, called the 'bandwidth', should continue to broaden with the switch from electronic to optical signal transmission; the speed of computation should increase as sluggish electrons are replaced as carriers of information by nimble particles of light; and the volume of computation may mushroom with the help of novel devices such as quantum computers. The clouds of information that surround us will thicken into opaqueness.

Just how pervasive all this information is destined to become in our lives can be guessed from the plans currently on the drawing boards. In response to the increasing demands of scientists and engineers, the World Wide Web has begun to be replaced by a much more complex and powerful World Wide Grid, which will in turn yield to something even more inclusive with a name like Omninet, or Hypergrid. Eventually, all our artefacts may be linked into a network of networks of networks. At MIT a program called 'Oxygen' aims to make computation as ubiquitous as the air we breathe. It envisions computers built into our clothes, walls, furniture, cars and bodies, while a haze of 'smart dust', composed of microscopic computers, sensors and transmitters, wafts overhead. On the other side of the country, at Berkeley, a project called 'Endeavour' in honour of Captain James Cook's ship, is designed to create an ocean of data that will envelop people like fish in the sea. Electric rain will swell into a deluge.

Where are the limits? Challenged by Richard Feynman's observation that 'there's plenty of room at the bottom', physicists and computer scientists have tried to estimate just how much room there is, before they hit the walls imposed by the laws of physics; but it may be that, even before the uncertainty principle of quantum mechanics, the second law of thermodynamics and the relativistic speed limit inexorably make their powers felt, human influences will begin to put a brake on the expansion of the information universe and flatten its growth curve.

For one thing, we are learning that the impact of the information age is not as universal as it may seem. To us in the affluent West information technology appears to dominate life, but to the majority of the global population it is still largely irrelevant. The World Wide Web will not solve the problems of poverty when half the people in the world have yet to make or receive a phone call. Self-guided automobiles will not improve the standard of living of the three billion people who survive on less than $2 per day. Robotic surgery will not heal the more than one and a half billion who lack access to clean drinking water. Eventually an appreciation of the treacherous depth and width of the digital divide may begin to dampen our boundless appetite for information.

The real cost of information technology has not been generally appreciated either. A recent assessment suggests that the manufacture of a single 2 gram computer chip consumes thirty-six times its weight in chemicals, 800 times its weight in fuel, and 1600 times its weight in water. Nobody knows when this hidden cost will become prohibitive.

Additional constraints on the growth of the information universe will probably be set by human frailty – for both our propensity to make mistakes and the limitations of our puny brains are sure to make themselves felt. Data will indeed flow faster and more copiously through the pipelines of the future, but the information carried by the zeroes and ones will inevitably suffer degradation through human error. The problem is already beginning to become apparent. A large portion of what can be found – if it is found at all – on the Internet turns out to be garbled, badly organized, or just plain wrong. Consequently much of it is neither accessible nor useful.

In 1997 Nobel laureate Murray Gell-Mann, the chief architect of the quark theory of matter, was asked to reflect on the next fifty years of computing and its effects on everyday life. He responded with an essay entitled 'Pulling Diamonds from the Clay', in which he addressed the problem of finding meaning in a flood of data, and wisdom in an ocean of information. A technocrat at heart, he placed his hopes in the ability of science and technology to help solve the problems they have created for themselves by growing so big so fast. Nevertheless, his doubts about the success of that enterprise are suggested by the quotation from Edna St Vincent Millay with which he concluded his article:

Upon this gifted age, in this dark hour,
Falls from the sky a meteoric shower
Of facts ... they lie unquestioned, uncombined.

Wisdom enough to leach us of our ill
Is daily spun: but there exists no loom
To weave it into fabric...

If it is true that the limits on information-processing will turn out to be more human than physical, technological or economic, it is ironic that popular usage makes so much of the prefix 'cyber', as in cyber-café, cybersex, cybercrime, and cyberworld. 'Cyber' was introduced into the English language back in the 1950s by Norbert Wiener's word 'cybernetics', which referred to the science of control over systems. He derived it from the Greek *kybernetes* for helmsman or guide – whose initial K appears in the name of the American academic honour society ΦΒΚ, an acronym for the Greek maxim 'Philosophy, the guide of life'. As we surf the Web for the latest high-tech magic, it pays to remember the human roots of the word. A cybership without a human steersman is a vessel out of control.

For the present, though, information rains down upon us without respite. It consists not only of artificially produced information, cyberstuff, the product of computers and information technology. Another kind of precipitation, the equally copious rain of natural information, is even easier to catch. We don't have to bother holding out an antenna – we just need to open our eyes. Consider the scene around you. As you look around, what you see represents a great deal of information. Even though you are aware of an unbroken tangle of colours and shapes, there are several ways in which this image can be reduced to a data stream. One would be to look at it through a digital camera, a device designed precisely for translating images into zeroes and ones. Nature itself anticipated this technique in the human eye, which converts visual cues into electrical impulses that are differentiated in space and time, and can also be rendered in digital form. And before the picture reaches the eye, it is carried by individual photons, or atoms of light – another digital encoding. Even the brain, which consists of a vast neural network of cells exchanging electrical and chemical signals, turns out to be a powerful processor of information coded for the most part in the form of off-on signals, zeroes and ones. A more tedious rendering of the same scene would be achieved by a precise verbal description, written in English and translated into Morse code or its computer-age replacement ASCII – dots and dashes, zeroes and ones.

No matter how we analyse the process, our eyes record a wealth of information. And the same goes for our ears, noses, tongues and fingertips, which all send messages to the brain for processing and recording. Comparably prodigious quantities of information are manipulated not just in the brain, but in all living matter. Molecular biology paints the picture of cells as repositories of genetic information. The growth and behaviour of organisms is determined and controlled by the genetic code, the book of life that we have learned to store in supercomputers in the form of zeroes and ones.

Information abounds in the physical world; it seems to be woven into the very fabric of the universe. As humans, we not only acquire information through our senses, we also feel compelled to share it with each other. From the first cries of emerging humans that we hear echoed from a baby's crib, from the gossip of cave-dwellers around the communal fire to our satellite-transmitted e-mail messages, the appetite for information has been as integral to the human condition as the hunger for food and love. Indeed, the anthropologist Barbara King has even suggested that the human ability to exchange information through speech and gesture is not unique. It evolved, she believes, along with other traits we inherited from primates and should be seen as part of a continuum that extends from an amoeba's ability to extract information from its environment, through the dance of the honeybee and the song of a bird, to our modern methods of communication.

But if information, in its natural and artificial forms, is such an essential ingredient of the world around us, why is it not part of the vocabulary of physical science? Why does it not play a role similar to such useful concepts as space, time, mass, energy, atom, light and heat that have become indispensable for describing the material world? Why do we look in vain for 'information' in the index of Feynman's monumental textbook *Lectures on Physics*, which initiated two generations of physicists into the profession?

The answer, alas, is that, though fashionable, information is also a vague, ill-defined concept. Perhaps that's a virtue in the postmodern era, but to the physicist it poses a challenge. What, exactly, *is* information? Is it a scientifically useful idea? Can it be measured? Will it yield to mathematical analysis? Such questions are the stuff of this book.

2 The Spell of Democritus

Why information will transform physics

The gradual crystallization of the concept of information during the last hundred years contrasts sharply with the birth of the equally abstract quantity called energy in the middle of the nineteenth century. Then, in the brief span of twenty years, energy was invented, defined and established as a cornerstone, first of physics, then of all science. We don't know what energy *is*, any more than we know what information is, but as a now robust scientific concept we can describe it in precise mathematical terms, and as a commodity we can measure, market, regulate and tax it.

Although information, too, is beginning to be marketed and regulated, it is distinguished from energy, and made much more difficult to define, by the aura of subjectivity that surrounds it. Energy is located strictly within a physical system – metabolic energy in a jelly doughnut, electrical energy in a mobile-phone battery, chemical energy in a gas tank, kinetic energy in every puff of wind. Information, on the other hand, resides partly in the mind. A coded message, for example, might represent gibberish to one person, and valuable information to another. Consider the number 14159265 ... Depending on your prior knowledge, or lack thereof, it is either a meaningless, random sequence of digits, or else the fractional part of pi, an important piece of scientific information. The smell of subjectivity, of dependence on a state of mind, is the source of both the elusiveness and the power of the concept of information.

'Stop right here,' I hear my colleagues grumble. 'You don't want to go there!'

At the mention of the word subjectivity, physicists cringe. For more than two thousand years we have been striving for objectivity, systematically trying to banish all traces of personal opinion and feeling from our discourse. So we bristle when postmodern critics try to portray science as a social construct, and physics as an ingenious narrative with the same claim to truth as a Hopi creation myth. Physics is fundamentally objective, we insist. Mathematics, that most rational of all forms of human communication, is its language. Let artists, psychologists and philosophers celebrate and analyse personal impressions and reactions; we physicists will stick to the facts, thank you. If information is subjective, its role in the vocabulary of physics is compromised right off the bat.

The scientific earthquakes of the twentieth century have shaken our faith in the possibility of a completely objective grasp of the inanimate world. First, Einstein's theory of relativity revealed that no phenomenon – not even the flow of time itself – can be described without reference to the state of motion of the observer. An accurate clock read by an astronaut in a speeding rocket ship would, for example, run more slowly than its counterpart back at the space port. It's not just that the clock would be *perceived* to run more slowly, the way distant aeroplanes seem to crawl through the sky, it really *would* tick at a reduced rate. Time itself would slow down in the rocket, dragging with it the astronaut's pulse and rate of hair loss. While this assertion, which is eerily counterintuitive but undeniably factual, does not really render time less objective, it unavoidably inserts the role of the observer into the language of physics.

If relativity brought us to the edge of the abyss of subjectivity, quantum mechanics plunged us right in. According to the prevailing interpretation of the quantum theory, the state of a physical system, such as an atom, is not definite, the way the direction of a pointer is always a fixed quantity (although it may be unknown). Before an objective statement can be made about a particular atom, the observer must set up an experiment to measure some property – say, the atom's speed. The choice of apparatus freely made by a human being partially determines the possible answers to the question at hand. If a different apparatus were employed – to measure the atom's position, for instance – the precise value of that atom's speed would cease to have meaning. The speed would become not merely *unknown*, but fundamentally *unknowable*, like the taste of darkness or the colour of hope. The speed of an atom does not exist until the question is properly formulated, and the appropriate measurement is performed.

This essential observer-dependence of quantum mechanics lies at the root of what has been called 'quantum weirdness' precisely because it violates the vaunted objectivity of the scientific enterprise. It's as if reality itself were shaped by the questions we choose to pose – with seemingly contradictory answers emerging for different choices. Ours is therefore a 'participatory universe', in which, according to John Wheeler, the observer cannot be redacted out of the picture. Can information, with its own built-in element of subjectivity, come to the rescue, as Wheeler hopes, and shed light on the mysteries of quantum mechanics?

Physics deals primarily with the world of things 'out there', a world represented most memorably by the rock whose palpable reality the lexicographer Dr Samuel Johnson demonstrated with a swift kick. At the same time it is obvious that the mathematical description of the world originates and operates not in matter, but in the human mind. In view of this inescapable commingling of the objective with the subjective, where should we look for the source of that alleged clean split between the observer and the observed? Whence sprang the conviction that we can study the world the way tourists observe apes in the zoo, with a sturdy fence between them to give the impression that the realms of people and of apes, of the watchers and of the watched, are separate and independent? Through a lucky accident of history it happens that the birth of this momentous blunder can be located with precision.

About 2400 years ago the philosopher and mathematician Democritus of Abdera enunciated unequivocally that the world, in spite of appearances to the contrary, is made of atoms. Only a few fragments of his works survive, but a portion of one of those is remarkable for its succinctness and unparalleled historical significance. It is arguably the fundamental axiom of physics: 'Sweet is by convention, bitter by convention, hot by convention, cold by convention, colour by convention; in truth there are but atoms and the void.' Convention, for Democritus, was collective human opinion as opposed to reality. Convention was subjective; the truth objective.

Throughout the centuries physicists have followed Democritus in separating objective reality from subjective perception. To be sure, they have succeeded in moving both temperature and colour from the subjective to the objective column of the ledger by linking them to real physical properties of atoms, but those insights only underscore Democritus's dichotomy. As our understanding of the nature of atoms has grown over the centuries, the entries in both the subjective and the objective categories have changed without compromising the original duality.

The history of atomism is marked by ups and downs, and even by long periods of silence, but the atomic doctrine preached by Democritus never entirely disappeared before it reached its present commanding position in science. The notion of simple, irreducible and eminently comprehensible objects – atoms – existing 'out there', at the base of physical reality, unsullied by the messy complications of the world of perceptions, was just too neat to be passed up. Atomism has served as a guiding beacon for physicists and chemists, and ultimately rewarded them with power over nature on a level that Democritus could not have dared to imagine. His spell – the dream of objectivity – has held them firmly in thrall for more than two millennia.

Nature, however, does not respect human categories. In the twentieth century physicists finally learned by experience that the fence between themselves and the inanimate world is permeable, and that in many circumstances – at the atomic scale of distances, for example – it dissolves altogether. Absolute objectivity may be an illusion, albeit a convenient and enormously useful one. It has served us well and made physics the envy of those sciences that are forced to grapple with human reactions; but now we must go beyond it, and acknowledge that the interactions between observer and observed cannot always be ignored.

Ironically, it was Democritus himself who pointed out the flaw in the claim of absolute objectivity. The fragment in which he grants the ultimate reality to atoms is actually a little debate between the intellect and the senses, and reads in full:

> Intellect: Sweet is by convention, and bitter by convention, hot by convention, cold by convention, colour by convention; in truth there are but atoms and the void.
> The Senses: Wretched mind, from us you are taking the evidence by which you would overthrow us? Your victory is your own fall.

Democritus understood that all empirical evidence in science is collected through the mediation of the senses. We learn about atoms by peering through scanning tunnelling microscopes, by translating their random motion, revealed to the touch as warmth, into thermometer readings, by converting invisible nuclear events into the audible clicks of Geiger counters. If we deny the reality of sense experiences, we destroy all evidence for the reality of atoms! After successfully suppressing the second half of Democritus's insight during all these centuries, we stub our toes against it in the very atomic realm that it opened up for us. We cannot pretend otherwise any longer: eventually observers, equipped with both senses and intellects, will have to be included in a complete description of the physical world. As Wheeler put it: 'No theory of physics that deals only with physics will ever explain physics. I believe that as we go on trying to understand the universe, we are at the same time trying to understand man ... The physical world is in some deep sense tied to the human being.'

But how do we get there from here? Starting with objective reality, consisting of the atoms and the void, which we pride ourselves on understanding, and proceeding in the direction of human senses and consciousness, which are still shrouded in mystery, we begin by asking the question: What connects the two? What mediates between the atom and the brain? What agency originates in the atom, or, for that matter, anywhere in the material world, and ends up shaping our understanding of it?

We can find a hint in a modern rendering of the doctrine of

Democritus. In his celebrated *Lectures*, Richard Feynman boldly attempted to compress the entire scientific enterprise into a single sentence:

> If, in some cataclysm, all of scientific knowledge were to be destroyed, and only one sentence passed on to the next generations of creatures, what statement would contain the most information in the fewest words? I believe it is the *atomic hypothesis* (or the atomic *fact*, or whatever you wish to call it) that *all things are made of atoms – little particles that move around in perpetual motion, attracting each other when they are a little distance apart, but repelling upon being squeezed together*. In that one sentence, you will see, there is an *enormous* amount of information about the world, if just a little imagination and thinking are applied.

Apart from some details discovered in the intervening millennia, such as the nature of the forces between atoms, the message in (Feynman's) italics is identical to Democritus's fragment; but although both bluntly proclaim the fundamental importance of the atom, Feynman has added a new concept that expands upon this objective assertion. The word he uses twice in the brief passage is 'information'. What a burden he places on that innocent little term. He implies that the atomic hypothesis is somehow packed like an overstuffed suitcase with information about the world – so much that it will take his thick book, plus a little imagination and thinking, to unpack it.

With characteristic insight, Feynman here emphasizes a property of information that links it closely with physics, and indeed with all of science: the ability to compress data. By way of illustration, consider the orbit of Mars. In the beginning there were fat tomes containing lists of Mars's celestial positions meticulously measured and diligently tabulated. In time, these were replaced by Kepler's laws, simple algebraic formulas that contained in their tiny frames an infinite number of possible observations. But even Kepler's laws were not sufficiently succinct. They were in turn superseded by Newton's, and later Einstein's, laws of motion and of gravity, which contain not only the orbits of all the planets ever found or awaiting discovery, but also the motions of their parent stars as well as those of footballs, tidal waves, galaxies and all other massive bodies in the universe.

In the same way the atomic hypothesis, which is couched in words rather than mathematical symbols, also compresses a world of data about the most diverse material systems imaginable. No matter whether it goes by the name of generality, parsimony, economy or data compression, brevity of expression is an essential ingredient of science. It became part of the scientific method in the fourteenth century in the form of Ockham's razor, the philosophical principle that values simpler, briefer explanations over more complex ones.

In communications theory, the idea of data-compression becomes explicit. An infinite string of digits, such as 01010101010 ..., which it would require an infinite quantity of paper to write out in full, can be compressed into the single command 'Repeat 01 without stopping.' If the string is considered as a message, the cost of transmission is thereby reduced from infinity to a trivially small value. This kind of elimination of redundancy from messages, which represents a form of electronic shorthand, is obviously of primary economic interest in the field of communications engineering.

The convergence of the aims of science and communications theory – to compress information to its smallest possible volume – led Feynman to link physics with information. The latter, his famous passage suggests, provides the connection between matter and mind that we are looking for. It is the strange, compressible stuff that flows out of a tangible object, be it an atom, a DNA molecule, a book or a piano, and, after a complex series of transformations involving the senses, lodges in the conscious brain. Information mediates between the material and the abstract, between the real and the ideal. If we can understand the nature of information, and incorporate it into our model of the physical world, we will have taken the first step along the road that leads from objective reality to our understanding of it. We will have broken the spell of Democritus. Then physics will truly enter the information age.

3   In-Formation

The roots of the concept

In order to understand information, we must define it; but in order to define it, we must first understand it. Where to start?

Defining words from scratch is like pulling yourself up by your own bootstraps. The paradox animates an anecdote related by the nineteenth-century Austrian physicist Ludwig Boltzmann, the first scientist to have an inkling of the potential role of the concept of information in science. In high school, he recalled later in life, he had the naive ambition to discover a philosophy in which every concept was clearly defined as it was introduced. Once, when he heard a philosophical work praised as uncommonly lucid (it may have been by Hume), he immediately asked his older bother to accompany him to the library to fetch it. Unfortunately, the book turned out to be available only in English, which Boltzmann didn't speak. 'No problem,' quipped his brother, who did not share Ludwig's idealism; 'if this book really achieves what you expect of it, the language in which it is written doesn't matter because every single word will be clearly defined before it is used.'

To help define 'information', a trip to the library is no more helpful than it was to young Ludwig. The shelves may well bend under the weight of dictionaries that contain the relevant entry and of books with titles that include the word, but none manages to provide a robust, satisfying definition. There is no choice but to roll up our sleeves and start from scratch.

Just like the word 'letter', which refers not only to a written message, but also to the alphabetical symbols that compose it, the word 'information' has two different senses. The colloquial usage, as in 'personal information' and 'directory information', refers to the meaning of a message of some sort. The technical sense, on the other hand, emphasizes the symbols used to transmit a message, whether they are letters, numbers or just the computer digits zero and one. In this second sense 'information technology' is that branch of engineering that focuses on storing, transmitting, displaying and processing symbols, irrespective of what they stand for.

Evidently, the two connotations of 'information' are closely intertwined. The meaning of a message arises out of the relationship of the individual symbols that make it up, just as the meaning of a letter emerges from the particular juxtaposition of its letters; but in spite of this obvious connection, the distinction between the colloquial meaning and the technical definition of information is profound.

The literary scholar and the scientist, perennial antagonists of the 'two cultures' debate, tend to come to the task of definition of concepts from opposite directions. Humanists typically start with an intuitive notion, flesh it out in the light of common usage and etymology, and propose a working definition. With the help of specific examples they then polish their draft until they feel that it catches the right sense. That's how dictionaries are made, and thus words gradually acquire fixed meanings. This haphazard process has led to a number of useful colloquial interpretations of the word 'information', but the persistent, nagging question 'What *is* information?' has not been answered.

Scientists, especially physicists, usually develop new concepts in a more pragmatic way. Since mathematics is the language of physics, and mathematics deals with numbers, an essential ingredient of every physical theory is measurement, the assignment of quantities to qualities. For this reason new terms are frequently introduced by way of recipes for measurement, called 'operational definitions', which require no real understanding of what it is that is being measured. That's how temperature started out, for example, which began its starring role in physics and chemistry around the year 1600 as something no more subtle than the quantity that is measured by a thermometer – a number read on a scale. As experience accumulated, readings on different types of thermometers were compared, and more was learned about the nature of heat from a variety of experiments, temperature was finally unmasked in the middle of the nineteenth century as a measure of the average speed of molecules. The journey from operational definition to an understanding of the real meaning of the word 'temperature' took a quarter of a millennium.

Information, too, has been defined operationally. Unfortunately this technical, bottom-up definition is very restricted, and hitherto bears little resemblance to any of the common, top-down definitions. Eventually the two definitions of information should converge, but that hasn't happened yet. When it does, we will finally know what information is; until then we have to make do with compromises.

In this chapter I will briefly examine the common, everyday word before turning for the rest of the book to the technical definition. Etymology offers a clue. 'Information', 'deformation', 'conformation', 'transformation', and 'reformation' obviously derive from 'formation', which, in turn, comes from 'form'. Information is therefore the infusion of form on some previously unformed entity, just as de-, con-, trans-, and re-formation refer to the undoing, copying, changing, and renewing of forms. Information refers to moulding or shaping a formless heap – imposing a form onto something. So the question of its meaning reverts to the more fundamental one: What is *form*?

The word 'form' entered Western philosophy as a translation of Plato's word *eidos*, the root of the words 'idea' and 'ideal'. Plato paints a picture of a world in which every object and attribute is but a pale, imperfect copy of a perfect, abstract ideal, a *form*, or archetype, which resides somewhere in an imaginary heaven. Thus a horse is but a copy of the form of horse-ness, the horse of horses, the *Ur*-horse, the ideal horse that has shed all material properties. Similarly, if you are good or beautiful, you are not really good or beautiful in a profound, ideal sense, you merely have some characteristics that reflect, in a crude manner accessible to our senses, the forms of goodness and beauty. As a boy I was intrigued by this untouchable world of essences, and, being more visually than verbally inclined, tried hard to imagine what the essence of horse-ness would look like, or how I would recognize a simpler form, such as that of a pencil. But even though, unlike Boltzmann, I had no older brother to tease me, I grew out of my youthful stage of Platonic idealism, and went on reasoning without forms.

Aristotle had trouble with Plato's forms too. What proof is there, he asked, that these things called 'forms' enjoy a separate existence? Instead of rejecting form altogether, though, he defined it as the sum total of the essential properties of a thing. An essential property of a horse, for example, is quadrupedalism, whereas colour, being variable and consequently accidental, is not part of its form. Two horses share the essence of horse-ness, Aristotle teaches, but there is no horse-ness without a real horse. In his theory of perception he assigns an important function to form. Our understanding of the material world, he claims, depends on having forms within our intellects: 'It is not the stone which is present in the soul, but its Form.' And these mental forms he calls ideas, abstractions, or concepts. Whatever traces of the great classical debate about the nature of form remain in our definition of information, they are more indebted to Aristotle than to Plato.

A much more common use of the word 'form' crops up in biology, where the infinity of shapes of living organisms provides us with a spectacle of awesome profligacy. The first modern biologist who tried to tame this profusion realized that the tool he required for the job, namely mathematics, would apply as well to other manifestations of form. In his book *On Growth and Form*, which appeared 1917 and inspired a small library of studies of symmetries and other patterns in nature, the Scottish naturalist D'Arcy Thompson wrote:

> We have learned ... that our own study of organic form ... is but a portion of that wider Science of Form which deals with the forms assumed by matter under all aspects and conditions, and, in a still

wider sense, with forms which are theoretically imaginable ... The mathematical definition of a 'form' has a quality of precision which ... is expressed in few words or in still briefer symbols, and these words and symbols are so pregnant with meaning that thought itself is economized; we are brought by means of it in touch with Galileo's aphorism (as old as Plato, as old as Pythagoras, as old perhaps as the wisdom of the Egyptians), that the Book of Nature is written in characters of Geometry.

The recognition that form must ultimately be expressed in mathematical terms carries over into the modern technical definition of information.

Thompson's idea of form could be expressed more clearly by the word 'shape', but it pays to take his advice and cast the net further out. The writer Paul Young, for example, collected eight synonyms for the word 'form' as he tried to pin down the nature of information: arrangement, configuration, order, organization, pattern, shape, structure, and relationship. Searching for the most general concept that can cover all possible applications in mathematics, physics, chemistry, biology and neuroscience, he settled on the term 'relationship' among the parts of a physical system, taking care to interpret the word in the broadest possible way.

D'Arcy Thompson's most famous examples of natural forms are those of the skeletons of marine micro-organisms called radiolaria. Alan Turing, the English genius whose contributions to the science of information ranged from fundamental theorems in logic to the cracking of the German army's Enigma code, worked for years on a mathematical theory of shape, with the goal of describing these exquisite forms. These structures exhibit patterns of such stunning beauty and intricacy that people who encounter them for the first time must be forgiven for doubting their reality. In fact, they resemble nothing so much as a certain kind of Chinese bone carving that tourists can't believe is made in one piece. The simpler examples resemble soccer balls with spikes, while others are graceful polyhedra with curved surfaces bulging outward. All of them astonish by the subtlety of the relationships between their parts.

The word 'shape' seems too paltry to describe such forms; if they had been secretly carved to incorporate coded messages, just imagine how much information they could transmit!

Complexity of form is not the only key to the amount of information a shape can hold. Take a far simpler, non-biological form – a circle, for example. Its essence resides in the relationship of its parts: each point on the periphery has the same distance from the centre. This kind of mathematical characterization, multiplied and complicated a millionfold, is capable of capturing all of nature's fabulous forms, at least approximately. Nevertheless, when we begin to quantify information, we will be surprised to learn that in principle, if not in practice, even an ordinary, unadorned circle can encode an infinite amount of information.

Further, the relationships of the parts of a system don't have to be spatial. Consider two rocks in outer space, and ignore their shapes and appearances. The relationship between them is summarized by the distance between them, and their mutual attraction as described by Newton's law of gravity. Since they move toward each other, the distance between them changes in time, so it would seem that their relationship cannot be expressed in purely spatial terms. Nevertheless, it was to be recast by Einstein in the general theory of relativity in the language of geometry – albeit in a space of four dimensions. So powerful is the human visual apparatus that physicists consistently strive to picture their equations. 'Geometrizing a theory', they call this exercise, and value it highly.

Other non-spatial relationships are logical or causal – A plus B equals C – and are exemplified by neural connections in the brain and electronic pathways in a computer. Both logical and causal relationships, if we adopt Young's definition, are forms. Temporal relationships such as the pattern of electrical pulses tapped out by a telegrapher, are, of course, paradigmatic of information transmission. The alphabet itself is a more sophisticated method for transmitting messages than this, each character being represented by a single letter; but again, most letters, by themselves, mean nothing in isolation. Literature resides in the pattern, in the way the letters are strung together, in their relationships to each other.

Tonal, temporal and energetic relationships combine in musical forms, spatial and colour relationships in paintings.

The further we pursue this string of associations, the more the world appears to be a tangled web of relationships. Relations are the stuff of science. 'The aim of science is not things in themselves, as the dogmatists in their simplicity imagine,' wrote the French mathematician Henri Poincaré, 'but the relations between things; outside those relations there is no reality knowable.' The Austrian philosopher Ludwig Wittgenstein pushed the point to its logical conclusion: 'We cannot think of any object apart from the possibility of its connection with other things.' This bothered him terribly, because he shared with the other Ludwig, his countryman Boltzmann, a yearning for a completely defined and consistent philosophical system. In the end he was stymied: it is impossible to define anything without first defining other things.

In human life, Wittgenstein's insight is almost self-evident. From birth we are enmeshed in a web of relationships that defines us and to a large extent determines our destiny. No one has expressed this truth more graphically than the Italian novelist Italo Calvino. His novel *Invisible Cities* is a haunting fantasy of travels through imaginary cities that symbolize different aspects of the human condition. In *Ersilia* the townspeople connect each others' houses with colour-coded strings to indicate relationships: '... white or black or grey or black-and-white according to whether they mark a relationship of blood, of trade, authority, agency'. Every relationship is represented, every house festooned with an ever-growing number of strings. Eventually, the tangle becomes so thick that movement becomes impossible, and the residents are forced to leave. The last sentence of the story is a poetic definition of the nature of community: 'Thus, when traveling in the territory of Ersilia, you come upon the ruins of abandoned cities, without the walls which do not last, without the bones of the dead which the wind rolls away: spiderwebs of intricate relationships seeking a form.'

Form, in short, expresses relationships, and this insight carries over into the concept of information.

Information, however, is not the same thing as form. The tiles on my bathroom floor display an interesting pattern, which has a form, and might even be called 'a form', yet there is no useful sense in which that pattern could be referred to as information.

Information carries a connotation of activity that is absent from mere form. Cicero used the verb 'inform' to signify giving shape to something, forming an idea, and moulding a person or his mind. Art critics often use the term in sentences like this: 'Picasso's personal experiences in the war inform all his paintings of that period.' Information, then, refers to imposing, detecting or communicating a form. It connotes change. It possesses 'informative power', in the words of science writer Robert Wright. Indeed, when we think about information, we associate it with learning something we didn't know before, with the news (which the French call *les informations*), with some kind of transfer of knowledge. To be sure, information can be stored in a memory, but that's a temporary arrangement. If it were permanently locked up with no possibility of flowing in or out, it would be useless, and not really information at all.

Information is the transfer of form from one medium to another. In the context of human information-exchange, 'communication' is a more descriptive term than 'transfer', and since form is about relationships, we can tentatively define information as the *communication of relationships*.

Two final remarks are necessary to round out this definition. The first concerns the way a form is imprinted on a medium of transmission. Translating a pattern in one medium into the same pattern, expressed differently in another medium, is called coding, and the dictionary that accomplishes it is the code. (Contrary to popular usage, a code is usually not secret. Secret codes, the concern of cryptographers, are called ciphers.) By way of example, consider listening to a live piano recital on the radio. The music, whose pattern is stored as notes on paper, or in some mysterious fashion in the player's brain, is translated into physical gestures, and from there into vibrations of the instrument. These, in turn, excite a sequence of pressure waves in the air, which induce vibrations in a distant microphone. That device delivers a coded electrical signal to a transmitter, which causes an oscillating current to slosh back and forth in a transmitting tower. This current is accompanied by electromagnetic radiation in the form of radio waves, electric rain pelting your receiving antenna. Then the transformation is reversed, step by step, until your brain registers the music. The code has changed many times in the intervening process, and none of the individual signals, be they mechanical, acoustical, electrical or nervous, resemble their counterparts in other media. At the same time, the kind of energy that was associated with the signal also changed with each successive encoding of the musical information; but despite this, the pattern, the relationship of the magnitudes and positions in time of the individual signals – in short, the form of the music – has survived intact throughout.

The second explanatory remark will sound perverse in view of what has preceded it. Even though information is, as Young put it, a flow of form, through most of this book we will consider information as a static phenomenon. The method of suspending time is trivial: a changing pattern can be recorded as a graph in which time is represented by distance along the horizontal axis. Thus a telegraphic message, which in reality consists of a temporal sequence of dots and dashes produced at one point in space, is most easily thought of as a long tape on which the dots, dashes and pauses are recorded in ink. More dramatically, an entire movie, with its complex sounds and actions, is encoded on a strip of magnetic tape in the form of dots and blanks, ones and zeroes. Thus, no sooner is time introduced as an essential factor in the definition of information than it is yanked off the stage again by a mechanical trick.

In-formation – the infusion of form – the flow of relationships – the communication of messages. The halting explanations by philosophers such as Aristotle, scientists such as D'Arcy Thompson, and writers such as Young and Wright, superimposed on our own preconceptions, help to endow the word 'information' with meaning. No crisp, robust definition emerges, but every gloss brings us closer to an understanding of that elusive concept; and as we turn from the rich, layered humanistic version to the skeletal technical definition, we must keep in mind that in the end the two must be reconciled.

The scientific measure of information

In contrast to the vague verbal definition of information, the technical definition, though skeletal, is a model of specificity and succinctness. Claude Shannon, the founder of information theory, invented a way to measure 'the amount of information' in a message without defining the word 'information' itself, nor even addressing the question of the meaning of the message. He produced, in effect, an operational definition like that of temperature, except that his measuring device – a simple recipe – is not a physical apparatus, like a thermometer, but a conceptual tool. The hope is that Shannon's definition will make contact with the meaning of information before a quarter of a millennium has passed, as it did between the invention of the thermometer and the realization of just what it is that a thermometer measures.

Shannon's information measure is most easily applied to a message that consists of a string of binary possibilities – yes or no, heads or tails, zero or one – each of which is equally likely at every step along the string. According to Shannon, each choice corresponds to one bit (short for 'binary digit') of information. Communicating to a friend the outcome of three consecutive tosses of a penny, for example, requires three bits of information, which would define eight possible strings of heads and tails. More generally, Shannon's recipe is simple: *To find the information content of any message, translate the message into the binary code of the computer and count the digits of the resulting string of zeroes and ones.* The number so obtained is called 'Shannon information', and the technique is known, somewhat dismissively, as bit-counting.

Shannon liked to illustrate the power of bits by describing how the game of Twenty Questions, if played cleverly, can deal with an enormous amount of information. The trick is to divide the possible answers into two roughly equal batches, over and over again. As long as each question is designed to differentiate between more or less equally probable alternatives, each answer reveals one bit. In trying to guess the location of a mystery town in the continental US, for example, ask either: 'Is it east of the Mississippi?' or: 'Is it west of the Mississippi?', because both questions neatly divide the country into two equal areas. The next question might be: 'Is it north of the Mason-Dixon line?', because that boundary, which separates Maryland from Pennsylvania, also bisects the country. On the other hand, 'Is it Camden, NJ?' is a bad question, because the answer 'No' would not eliminate much of the country, and would therefore adduce very little new information. (Of course, the answer 'Yes' would instantly convey the full information in this case, but when you're calculating probabilities, you can't assume a lucky strike, like a royal flush off the deal in poker. You have to work your way through *all* the possibilities and eliminate them systematically.) With good strategy, twenty answers elicit 20 bits of information, which in turn correspond to a single choice among 1,048,576 equally probable alternatives – a number computed by multiplying together twenty factors of two. That should suffice to identify any town in America.

The trouble with Shannon's nifty operational definition is that it says nothing about the intended meaning of the message. A telegram in Morse code consisting of forty dots and dashes would contain exactly the same information content of forty bits, whether it spelt out 'happy birthday' or 'wjirdkuierntuc'. Worse, a second copy of that same telegram would double the available Shannon information without producing any new knowledge. The precision of the definition has been bought at the price of ignoring the meaning. In the exercise of bit-counting, Wheeler's really big question WHAT MAKES MEANING? is not even asked. This glaring deficiency of Shannon's definition appears to strike a blow for the humanistic side of the 'two cultures' debate. Scientists, it might seem, address the quantitative but superficial aspects of the thorny problem of information, while ignoring the qualitative, more interesting ones.

This criticism ignores the lessons of history, however. The French biologist François Jacob, who won a Nobel prize for helping to discover the role of RNA as the messenger molecule for genetic information, pointed out: 'The beginning of modern science can be dated from the time when such general questions as "How was the Universe created? ... What is the essence of Life?" were replaced by more modest questions like "How does a stone fall? How does water flow in a tube?" ... While asking general questions led to very limited answers, asking limited questions turned out to provide more and more general answers.' It is the passionate pursuit of particular puzzles, not the pondering of profound problems, that leads to universal insights. The history of physics thus raises the hope that Shannon's simple recipe for quantifying information, which underlies the eminently successful science of information theory, will, in time, shed light on the deep question: What is information?

Following Jacob's advice, most of this book will deal with the technical, operational definition of information – the number of bits in a message – together with its extension to quantum information, and the lessons we can learn about reality from both of these. These are modest questions, but if we push them to their limits in accordance with the scientific method, they will eventually lead to real understanding.

Shannon's definition singles out the binary code for the purpose of bit-counting. Why? Why not stick with the conventional decimal system that every preschooler learns? Would it not be just as convenient, and more familiar, to assign decimal equivalents to all alphanumeric symbols, and *then* count the digits?

Far from selecting the binary code arbitrarily, or believing it to be the simplest possible scheme, Shannon proved that it is, in fact, the least expensive way to handle information. It uses up the smallest amount of resources in the form of electronic memories and bandwidth in communication channels. Nature, in turn, often imitates art by picking the cheapest, most energy-efficient solution when it has a choice. (Remember the bees: Darwin himself pointed out that the hexagonal arrangement of their hives supplies the maximum amount of living space at the lowest cost in building material.) The binary code, therefore, recommends itself as the ideal tool for measuring information in technology as well as in nature.

To get a flavour of Shannon's proof, consider a sailor who wants to signal a number between 0 and 127 by means of flags. If he decides to fly just a single flag to do the job, he needs to have 128 different flags in his locker. A less expensive strategy would be to fly three flags to spell out the number in the decimal system. To do this, he would need to own just twenty-one flags – ten for the units, ten for the tens and one for hundreds. The cheapest technique is based on the binary system. With only fourteen flags – seven zeroes and seven ones – he can compose any number up to 127. The price of this method – fourteen flags in place of 128 – represents a cost saving of some 89 per cent.

By considerations of this kind, Shannon was able to show not only that the binary code is the most efficient way to store digital data, but also that digital data processing is more efficient than analogue technology. A familiar illustration of this discovery is found in the reproduction of music. Whereas a long-playing record stored half an hour of music by means of a jiggling groove on a disc the size of a dinner plate, a CD can store four hours in the form of zeros and ones – in $\frac{1}{25}$ the space. Digital is the way to go, and bit-counting the natural measure of the volume of data.

What about the meaning of the information being measured? Does it matter whether you store music or noise? Shannon's measure is silent on the subject, as he pointed out in the second paragraph of his famous 1948 article. 'Frequently the messages have *meaning*,' he wrote, and used the adjective 'semantic' for 'meaningful'. Then he continued bluntly: '[The] semantic aspects of communication are irrelevant to the engineering problem.' In other words: forget the meaning.

A year later the applied mathematician Warren Weaver amplified this remark in a book he wrote together with Shannon. Weaver identified three levels of complexity in dealing with the problems of information and its communication, of which the first and lowest level begins with the question: 'How accurately can information be transmitted?' Bit-counting allows a simple numerical answer: just measure the amount of correct information, symbol by symbol, of the message at the receiver end, and divide it by the amount of information transmitted. The result, expressed as a percentage, provides a quantitative measure of the faithfulness of communication.

Much of the effort expended on information technology in the last half-century has gone into the design of machinery that achieves values of this measure that are unimaginably close to 100 per cent, and still increasing. To appreciate the magnitude of this feat, just imagine dictating to a friend the string of ten million zeroes and ones that make up a single colour snapshot sent over the Internet. How many errors do you think you'd make? How many lines would you mix up? How many zeroes or ones would drown in the sea of bits that surround them? Yet the World Wide Web handles trillions of pixels, billions of pictures, millions of movies – all day long, every day – with astonishing reliability.

Weaver's second and third levels of information concern the semantic problem and, related to it, the effectiveness problem – neither of which is addressed by bit-counting. The semantic problem is: 'How precisely do the symbols convey the desired meaning?' This question plunges straight into the abyss of metaphysical inquiry, which physicists are loath to plumb. (Robert Wright, who coined the phrase 'informative power' mentioned in the last chapter, cautions: 'If we are smart, we will doggedly resist any impulse to think closely about meaning. On the other hand, if we were smart, we probably would not have gotten bogged down in the contemplation of information in the first place.')

On the third level we encounter the effectiveness problem. In the context of communications, the question is: 'How effectively does the received meaning affect behaviour in the desired way?' If

I receive a message, and understand it perfectly, but interpret it in a way that is different from what the sender of the message intended, then the message somehow did not effectively carry the intended meaning. If someone shouts 'Fire!' for example, I may hear the word correctly (success on level 1), understand the English meaning of the word (success on level 2), but helpfully click my lighter instead of running out of the building (failure on level 3).

The effectiveness problem is related to the usefulness of information. There is a hint here that, unless information leads to significant consequences, it is not really information at all, and should not be counted. For this reason the physicist Murray Gell-Mann, together with his colleague Jim Hartle, has introduced a formalized substitute for the infamous and enigmatic 'observer' of conventional quantum theory, which they call an IGUS, for information gathering and utilizing system. Gell-Mann explains the term: 'In quantum mechanics there's been a huge amount of mystical nonsense written about the role of the observer. Nevertheless, there is some role for the observer in quantum mechanics, because quantum mechanics gives probabilities. And you want to talk about something that is betting on those probabilities. That's the real role of the observer.' In fundamental physics, in other words, it is not enough to transfer information from object to subject. Information gathering by itself, without observable effects on the gatherer's behaviour, is a pointless pursuit. By the same token, an IGUS, which *does* react to the information, is a far more complex and sophisticated being than a mere machine.

In spite of its shortcomings, bit-counting, which sits at the bottom of the hierarchy of the problems of information, is a potentially powerful tool. Adroitly wielded, it is a key that will open the way into higher levels of meaning. But whether or not information, analysed on any level, will be of use in physics is a different question. Its links with subjectivity may pose a problem, but may also turn out to be its greatest strength. There may be even more fundamental hurdles. Information, like energy and entropy, is an invented term; there is no guarantee that it will serve a scientific purpose. Is it too broad, perhaps, too vague, too intangible to be of use? Will it turn out to be simply too abstract to compete with such eminently tangible concepts as matter, force and motion?

5 Abstraction

Beyond concrete reality

The Swiss psychologist Jean Piaget identified four stages of human mental development: the sensorimotor stage (0–2 years) of discovering the self as part of the environment; the pre-operational stage (2–6) in which symbols and numbers begin to be used; the concrete operational stage (6–12), in which logic appears in the classification of ideas and the understanding of time and number; and the formal operational stage (12–adulthood), which culminates in the manipulation of abstract concepts.

As an example of this sequence, consider how a child learns to manage money. At first, it's just a matter of watching Mummy and Daddy in the shop, and noticing the routine followed at the checkout counter. Later, numbers are attached to currency, and the system of notes and coins begins to make sense. The pre-teen learns the logic of rendering payment and receiving change, the relative values of different goods and services, and the cause-and-effect relationship between duration of work and earnings. The teenager, finally, is ready to come to grips with such complex abstractions as debt, interest and the meaning of the word 'mortgage'.

The same pattern of 'observation → measurement → model or formula → law or abstract principle' is followed in learning to understand everything from money to higher mathematics. It is also a recurring theme in the history of science.

Look, for example, at the theory of planetary motion, which served as an example of data compression in chapter 2. Early humans simply noted the movement of the sun and the planets, and then correlated them with the daily, seasonal and yearly cycles that imposed their rhythms on personal and communal life. (This is the sensorimotor stage.) When numbers were invented and careful measurements were performed, the celestial movements were translated into elaborate lists of astronomical positions (pre-operational). Later, theories of the solar system in terms of concrete models, such as crystal spheres, accounted for recurring regularities, and were replaced by geometrical figures – first circles and later ellipses – which, though idealized, are still directly accessible to the senses (concrete operational). Only in the most recent times, however – since Newton and Einstein – have planetary data been compressed by means of such eminently abstract and ultimately unimaginable concepts as the gravitational force and the gravitational field (formal operational). Abstraction grew at every step, allowing ever-increasing generality and breadth of scope at the price of palpability and concreteness. The intellectual road from my grandson Teo's fascination with dropping a ball into a bucket to his understanding of the force of gravity recapitulates the journey from a caveman's observation of a rock's plunge off a cliff to Newton's law of universal gravitation. Increased abstraction is the hallmark of growing maturity.

From a broader perspective, all of classical physics followed the same sequence. It began with ancient observations, grew into Greek astronomy, flourished with the Newtonian synthesis of the seventeenth century, and culminated in the nineteenth century with the establishment of the laws of mechanics, heat, electricity and magnetism in the form of succinct mathematical equations. The philosopher Ernst Cassirer captures the trend toward abstraction with the remark: 'In its general structure, nineteenth century physics might be characterized as a physics not of images and models but of principles.'

In modern times, this tendency has continued to make itself felt with increasing intensity. It contributes to the reputation for difficulty from which physics suffers in the popular imagination. Art, architecture, music and literature, in all their forms, are supposed to be everybody's business, in part because they appeal directly to the senses, whereas theoretical physics and mathematics appear increasingly 'pulled away from' ordinary experience – the word 'abstract' comes from the Latin *trahere*, to pull, as in 'tractor'. The inclusion of information in the tool kit of physics, and eventually in the language of popular science, is part of the relentless trend toward abstraction that characterizes the evolution of modern science as much as the development of thinking in children.

The best example of the way physics rushes into abstraction is provided by the story of matter. Two and a half millennia ago the atomic hypothesis was a bold step away from common, sensory experiences into an inaccessible world of ideas. Since then, atoms have managed to float up out of the pit of metaphysics into our ken, and today it is possible to see, touch, manipulate and mould them. From products of philosophical speculation they have been transformed into the concrete building blocks of the universe. In terms of Piaget's categories this progression appears to be retrograde – from invisible, abstract concepts to concrete objects – but that appearance is misleading, for even as atoms became more real, they dissolved. This catastrophe has been called a 'flight from common sense' by the Nobel laureate Philip Anderson. The turning point came with Ernest Rutherford's discovery of the atomic nucleus in 1913, when the miniature pellets of Greek atomism suddenly turned out to consist mostly of empty space. But worse was yet to come. Even their constituents, the electrons, protons and neutrons that make up the modern atom, and that were originally thought of as particles with the same properties as grains of sand, lost their solidity. Science has taught us that what we see and touch is not what is really there.

Particles, we now know, share many properties with waves. Accordingly the hybrid concept of *matter waves* was introduced – a cross, as the discoverer of the electron put it, between a tiger and a shark – each creature dominant in its element but helpless in the other's. Since both particles and waves are familiar, this reconciliation of irreconcilables is at least as imaginable as some exotic offspring of a tiger and a shark. Unfortunately even that's not weird enough. Even matter waves can't rise to the challenge of describing atomic reality.

Instead, electrons in atoms are described by a mathematical device called the Schrödinger wave function, which (except in the special case of hydrogen, the simplest of all atoms) does not exist in the familiar three-dimensional world. It lives on paper and in the mind (or the soul, as Aristotle would have said), and does not make contact with actual measurements in the laboratory until it has undergone a set of rigorously prescribed mathematical manipulations. Even then it does not describe the palpable electron cloud surrounding the atomic nucleus, as its inventor initially believed. In fact, it does not describe anything that actually is, only the probability of what might be. The wave function could easily be dismissed as too anaemic, too abstract, to serve a fundamental ingredient of our world picture, if it were not for the stubborn fact that it works spectacularly well, and that nobody has discovered a more robust substitute.

If the wave function does not represent a faithful picture of an atom, what exactly is it? In different experiments one can measure the positions, the velocities, the energy levels, the magnetic effects and a variety of other properties of the electron. All these properties can be predicted from the wave function, which therefore acts as a succinct mathematical encoding of information resembling a map of potentiality – a catalogue of possibilities. If the wave function is nothing but a storehouse of information needed to make correct predictions, then the stuff of the world is really, at bottom, information.

But the story of matter does not end here. When the dictates of the special theory of relativity were imposed on quantum mechanics, particles, and even their associated wave functions, lost their effectiveness as components of the physical world picture. The successor to the wave function is arguably the most abstract concept in science. Called the *quantum field*, it is more fundamental than particles and more basic than forces. The root of the idea was Earth's familiar, though enigmatic, gravitational field, which gently tugs at us from birth, and in the end pulls us into the grave;

but the quantum field is stranger by far: it is utterly unimaginable.

A quick glance at the idea demonstrates that its outrageous abstraction – which is to say the utter lack of visual or verbal images it suggests – is not an impediment to its usefulness. If the quantum field, which has been around for more than half a century, can be assimilated into modern physics without provoking a public howl of anguish, or even a debate within the physics community, then the concept of information, no matter how abstract it will turn out to be, may also have a chance to establish itself.

In the modern view, the universe is filled with quantum fields, ethereal auras, one for electrons, one for each quark, one for every type of elementary particle. These fields coexist in space like milk poured into water. But unlike milk and water, which retain their identities at the molecular level, quantum fields are joined at each point in space-time by intimate bonds that allow them to exchange energy and even to transform into other fields. A particle – say an electron – is a ripple in the electron field. A proton, located elsewhere and carrying a different amount of energy, is a separate ripple in the proton field. These ripples can touch and scatter like waves on a lake, or they can coalesce, raise new ripples in a neutron field and a neutrino field, and in the process subside into oblivion.

A quantum field is a kind of Aristotelian form. Inscribed in its mathematical formulation are all the essential properties of a particle, such as mass, charge and magnetic properties, which are shared by every copy of that particle in the universe. Accidental properties, on the other hand, such as location, speed and direction of motion, vary from individual to individual. They characterize the ripples in the field, not the field itself. If the field is a form, the particle is its manifestation in the real world.

Steven Weinberg, Nobel laureate and major contributor to the creation of the Standard Model of Elementary Particles, used to believe that the world is made of fields, and nothing else. '*The essential reality is a set of fields,*' he wrote, 'subject to the rules of special relativity and quantum mechanics; all else is derived as a consequence of the quantum dynamics of those fields.' Where once Democritus had declared that only atoms and the void are real, Weinberg located ultimate reality (the elusive subject of John Wheeler's birthday symposium) in quantum fields. More recently, however, he has admitted that the multitude of different kinds of fields required for the Standard Model is too profligate. He now hopes that even more fundamental concepts than fields will eventually unify and simplify the physical world view, though it is not yet clear what these concepts will turn out to be. Ockham's razor, he trusts, will once again prune the proliferating set of primitive entities, and reduce the number of a priori assumptions that must be fed into the theory at the outset; but for now, practical calculations of real laboratory measurements are always couched in the language of fields.

Quantum field theory, the mathematical framework of particle physics, has scored magnificent successes not only in bringing order into the confusion of phenomena at the foundations of the material world, but also in reproducing quantitative measurements with impressive accuracy. Richard Feynman was fond of remarking that the quantum field theory he had helped to develop was capable of predicting the magnetic properties of electrons with a precision of one in a billion – a feat he compared to determining the distance between New York and Los Angeles to the width of a hair. The Oxford theorist Roger Penrose likes to top this boast by pointing out that the correct prediction of general relativity (a classical field theory) for the motion of the double neutron-star system PSR 1913 + 16 is a hundred thousand times more precise.

The fact that field theory has achieved all this without ever being called upon to address the question of what a field actually is neatly demonstrates that abstractness is no impediment to usefulness in theoretical physics. Of course, physicists would prefer their theories to suggest simple, intuitively appealing images for making sense of the inaccessible world of the quantum. Visualizability, or what the Germans call *Anschaulichkeit*, is a desirable attribute; but if it can't be achieved, if a theory satisfies the normal requirements of the scientific method – prediction of experimental measurements, unification of apparently disparate phenomena, connection with other theories – while retaining the forbidding level of abstractness that characterizes quantum field theory, then it just has to be accepted as it is. Where images and models fail, abstract principles must suffice.

The lifeblood of quantum fields – the stuff that animates them – is energy. If you want to find out how nature works, just follow the energy trail; but energy, like blood, is too formless, too undifferentiated; it carries fuel, not messages. Perhaps information, which is poised to become the next fundamental addition to the arsenal of theoretical physics, will fill that role. Like the quantum field itself, information has roots in the ancient theory of archetypes; it too deals primarily in relationships rather than in things, and it promises to be just as abstract.

For all the difficulty of its definition, however, information enjoys a decisive advantage over the quantum field as the primitive concept from which the world might be constructed. Unlike the unimaginable field, information is a common, everyday idea with immediate intuitive appeal. No matter what bells and whistles scientists eventually hang on it, its core will remain untouched. Ask a ten-year-old to describe a quantum field, and you will get a blank stare; but if you say you need information, you will be asked, 'Information about what?' We may not know how to define it, but we certainly recognize it when we see it. What better basis upon which to build a coherent picture of the material world than a concept that is thoroughly familiar to every child?

And not only the material world. As the study of life proceeds beyond anatomy, down the ladder of sizes and up the ladder of abstraction, it encounters along the way the science of genetics, which, remarkably, also deals primarily with information. This coincidence augurs well for the promise of information as a common currency for *all* the sciences.

6 The Book of Life

Genetic information

With singular historical appropriateness the year 1900 witnessed the birth of both quantum mechanics and molecular genetics, arguably the two outstanding achievements of twentieth-century science. On Monday, 6 June 2000, a century later, at a ceremony in the White House, the human genome, a record of the totality of human genetic information, was opened to the world like a vast book written in just four chemical letters. It contains an unimaginable wealth of data whose interpretation during the coming decades will be as much a problem of information-processing as of genetics. The way biology and physics, with their disparate aims and methods, converged in this hundred-year quest is a testament to the unity of science. From the beginning, the central issue was information – information about matter at its most fundamental and information about life itself.

The quantum revolution began on 14 December 1900, when Max Planck, a decidedly unrevolutionary professor of theoretical physics, reported a bizarre hypothesis to the Berlin Academy of Sciences. The problem he had been wrestling with for six years concerned the interpretation of experiments performed for the benefit of the German gas and electrical lighting industries, with, seemingly, very little fundamental significance for physics. No one, least of all Planck, had an inkling that it would undermine the view of the world that had been developing since the time of the Greek philosophers in the unbroken tradition called classical physics.

The experiments in question were refinements of the observation that an incandescent object, such as the heating element of a light bulb or of an electric range, emits light of various colours. When the metal is relatively cool, it looks reddish, because among the rainbow of colours it radiates, red hues predominate. Upon further heating, the element begins to appear white-hot as the intensities of the blue colours rise. By the autumn of 1900 German physicists had plotted precise curves of intensity versus wavelength – or colour – for bodies of various temperatures. Their experimental work completed, they turned their graphs over to the theoreticians for explanation. Planck found the challenge tougher than expected.

Combining thermodynamics, the science of heat, with the electromagnetic theory of light and of thermal radiation, he found himself unable to proceed without making a seemingly unphysical assumption – the quantum hypothesis – which he regarded as a mere temporary expedient. Thermodynamics was written in the language of atoms, which were regarded as discrete, granular entities. Radiation, on the other hand, was described as being continuous, like a smooth water wave or an undulating beach seen from a great distance. In order to reconcile these two radically different points of view, he assumed that matter emits energy not continuously, but in tiny discrete portions. He called these atoms of radiation *quanta*, the plural of the Latin *quantum*, meaning 'how much', as in the word 'quantity'. By regarding both electromagnetism and thermodynamics as granular, Planck was able to bring his calculation to a successful conclusion. The formula he derived fitted the experimental data with such spectacular precision that it earned him a Nobel prize in 1918.

Its foundations were suspect, however. Everyone, including Planck, believed that radiation was really perfectly continuous, and that quanta would eventually turn out to be superfluous artefacts, like the faint pencil lines draftsmen use to guide their drawings and erase at the end.

Everyone, that is, except Albert Einstein. In 1905 Einstein, then twenty-six years old and endowed with a prodigious physical intuition, boldly proposed that the quantum hypothesis was not just a convenient subterfuge, but a real and fundamental natural fact: light and radiant heat, he claimed, did indeed consist of bundles of energy (which later came to be called photons). For the next five years, the Planck/Einstein hypothesis did not command much attention in the physics community, but beginning around 1910 it became the basis of a new way of thinking about the world. Continuity turned out to be an illusion and discreteness ruled. Classical physics, at least in the atomic realm, was dead; but big as this jolt was, the one that shook biology proved, if anything, even more violent.

True, by the turn of the century, the observations of the Augustinian abbot Gregor Mendel were already a generation old, but the vehicle he had chosen in 1866 to communicate them to the world, the transactions of the Scientific Society of the City of Brünn, Austria (now Brno in the Czech Republic), had been far too obscure to carry his message to the wider scientific community. Accordingly, the laws of heredity languished unappreciated until the year 1900, when three biologists working independently in Amsterdam, Berlin and Vienna rediscovered them and brought Mendel the posthumous fame he deserves.

Mendel, who was born in 1822 to peasant parents, had been educated at a monastery before studying mathematics and science at the University of Vienna. When he failed his teacher's examination he returned to the monastery and began a series of patient, painstaking experiments intended to bring quantitative methods to the study of heredity. The garden pea, which breeds quickly and is easy to manipulate, had been the subject of many earlier investigations, so he, too, chose it as his experimental vehicle. Working with thirty-two different varieties, which he distinguished by observing seven different traits such as size and shape of various body parts, he grew and cross-bred many thousands of individual plants in search of regularities.

His most famous discovery concerned flower colour. When he crossed white-flowered peas with purple ones, the offspring were not, as one might expect, pale purple, but the entire crop came out looking like the purple parent. In the next generation of peas the surprise was even greater: three-quarters of the flowers turned out to be purple, and one-quarter pure white. That 3:1 ratio, which he documented with impressive reproducibility, great reliability and high accuracy, is now called the Mendelian ratio. It was typical of a host of other mathematical regularities he uncovered.

The implications of the laws were twofold. First, they brought the study of heredity into the compass of mathematical science, with which Mendel was comfortable and to which all scholarly disciplines yearn to belong. Secondly, that macroscopic regularities revealed a microscopic graininess: that fixed, persistent ratios of traits among populations result from fixed ratios for the individual members of the populations. Thus the almost exact 1:1 ratio of men to women among married people is easily explained by noting that each individual couple is constituted in that ratio, which automatically propagates to the whole group. A less obvious example is the fixed 2:1 ratio of hydrogen to oxygen in the composition of water, which led nineteenth-century chemists to the atomic hypothesis: that every water molecule is made of two atoms of hydrogen attached to one of oxygen. This explanation was remarkable for the date at which it was proposed – a century before the reality of atoms was universally accepted, and a hundred and eighty years before individual atoms became visible. That something as complex and mysterious as heredity can also be transmitted in quantized units must have come as a profound intellectual shock to Mendel and his disciples. Just as in physics, a granularity, a discreteness, turned up where none had been suspected before.

By 1911 the hypothetical unit of heredity had acquired the name 'gene'. Like atoms (the units of matter), and quanta (the units of energy), genes were discovered by inference at a time when their physical nature was utterly mysterious. In fact, the debate about whether genes were operational or material entities would simmer for decades. Nevertheless, their location in living matter was quickly ascertained. Karl Correns, the German member of the trio of biologists who rediscovered Mendel's work in 1900, suspected at once that the physical carriers of heredity must be chromosomes, the twig-like bits of stuff in the nuclei of living cells. And that wasn't all: even the positions of the genes on the chromosomes were soon discovered.

'The Lord God is subtle, but malicious he is not,' Albert Einstein said. Nature often presents us with seemingly unsolvable puzzles, but then provides nifty clues for their solution. In the case of genes, the hint was the observation that chromosomes happen to be linear, like tiny sticks, rather than flat, like pancakes, or lumpy, like potatoes. No one can say how genetics might have developed, or how heredity would work, if those had been nature's choices. The linear, one-dimensional structure of chromosomes provided the vital clue for the discovery of how genes line up on them.

The idea is simple: if you paint a red dot and a blue dot at arbitrary points along a stick, and then break the stick in two, the dots may end up either on the same fragment or on two different ones. That much is obvious. A little reflection prompts the additional remarks that the closer the two dots are to each other, the more likely they are to end up together, and conversely, the further they are apart, the more likely they are to end up on separate fragments. In particular, if the dots happen to reside at opposite ends of the original stick, they are certain to be separated by any break. This argument, which links the physical position of points on a stick to the numerical probability of their remaining together, is the key to the science of genetics.

Substitute genes related to different traits for red and blue dots, chromosomes for sticks, and a variety of cell divisions, recombinations and natural or induced mutations for breaks in the stick, and you get the basic logic of chromosomal genetics. When one individual inherits two different traits, it means that the two corresponding genes have stayed together, but when the traits appear in separate populations, we can infer that the genes have ended up on separate chromosomes that travelled separate routes. In 1913 a Columbia University student named Alfred Henry Sturtevant, in conversation with his teacher, the eminent geneticist Thomas Hunt Morgan, suddenly realized that such arguments would enable him to quantify the physical locations of genes. That same night he computed and drew the first primitive genetic map of a chromosome. The three traits whose relative positions he located were variants of characteristics of fruit flies: yellow instead of the normal grey body colour, white rather than red eye colour, and stunted wing length. From this embryonic chart the mighty human genome-mapping project grew.

By the 1930s biologists knew a lot about what genes did, and where they resided, but they still had no idea what they were made of. Genes obviously carried huge quantities of information from generation to generation, but how? Were they, perhaps, some kind of complex biological structures, miniature memories, too small to be visible but sufficiently potent to organize the growth and development of human beings?

To older geneticists, intent upon understanding the mechanisms of gene propagation, their physical structure was of secondary importance. As late as 1933 Morgan remarked in his Nobel lecture: 'It doesn't make the slightest difference whether the gene is a hypothetical unit or a physical unit.' But that indifference didn't extend to the younger generation. Morgan's student Hermann Muller, himself a Nobel laureate in later years, had seen the path ahead more clearly a decade earlier: 'Must we geneticists become bacteriologists, physiological chemists, and physicists, simultaneously with being zoologists and botanists? Let us hope so.'

His wish was fulfilled in the middle of the century, when the story of genetics switched from biology to physics. If genes were actual physical structures, their architecture and operation cried out for investigation. They were, after all, responsible not only for the physical properties of human beings, but also for our behaviour and feelings, to the extent that those are inherited. Understanding genes would go a long way towards understanding life itself. Nowhere is the intertwining of physics with biology in this endeavour described more eloquently than in the book *What is Life?*, by the co-discoverer of quantum mechanics Erwin Schrödinger. Based on lectures he gave in Dublin in February 1943, at the height of the Second World War, this little work of fewer than a hundred pages quickly became a classic, renowned for setting out the foundations of modern molecular genetics in simple, non-technical terms. It served as an inspiring intellectual catalyst for two generations of researchers.

In his inquiry into the material aspects of heredity, Schrödinger begins the way physicists habitually approach any new subject: he estimates a number. Numbers instil a feeling for the lie of the land, and furnish grist for the mathematical mill that is the physicist's principal tool. In a textbook example of consilience, the convergence of evidence from different branches of science on a common explanation, Schrödinger points out that the biological estimates of the approximate size of a gene agree with evidence based on physical evidence. Breeding experiments on plants and animals had led to detailed chromosomal maps, whose dimensions furnished the estimate that a gene normally contains no more than a million atoms. Direct observation of the lumpy structure of chromosomes by means of electron microscopes confirmed that estimate. While a million sounds like a huge number, on the atomic scale it is remarkably small: a cubical stack of a million atoms numbers only a hundred along each edge. Additional probing by means of X-rays, which can discern finer details than optical or electron microscopes, lowered the estimate even further to about a thousand or so, or ten to an edge when stacked up – a truly microscopic number that would turn out to be crucial. Notice that at this stage no mention is made of the identity of the atoms, or the way they are put together.

Next, Schrödinger asks: how long do genes retain their identity? The answer is easy – practically for ever. For hundreds of thousands of years, stretching into millions, humans have had ten fingers, two eyes and countless other traits. All this information is transmitted faithfully from generation to generation by genes. Except for rare, sudden, and dramatic mutations, genes are permanent, stable structures. If they were not, biology would be a tumultuous, unpredictable, lawless science, and the world a vastly different place.

Under the laws of classical physics, permanence and small numbers are incompatible. Smoothness, permanence, regularity can only emerge as statistical results from large numbers. Thus, the stability of a gold coin is made possible by the astronomical number of gold atoms it contains, and the predictability of the temperature of the sun by the multitude of hydrogen atoms that contribute to it. Small collections of atoms, analysed according to classical thermodynamics, are subject to continual statistical fluctuations, to jiggling and cavorting in an endless and disruptive dance that would soon destroy their pattern. If, by way of analogy, you carry half a barrel of water on the back of a pick-up truck, and the road is smooth, the barrel remains more or less half-full. But try the same with a coffee cup, and the fraction of the mug that remains filled becomes unpredictable. The surfaces of the water in both containers are jostled by similar vibrations, but in the bigger vessel they matter relatively less.

Schrödinger emphasized that, classically speaking, a solid block of a thousand atoms, or even a million, at the temperature of the human body, should be utterly incapable of encoding a message for an appreciable length of time, and certainly not for the aeons required to understand heredity. If it were cooled to the absolute zero of temperature, or near it, atomic motion would turn sluggish, and persistent patterns might endure, but at body temperature that option is not available. 'Permanence unexplainable by classical physics,' is his laconic conclusion. That leaves only one option. Since the only structures made of small numbers of atoms but endowed with a high degree of permanence are molecules, Schrödinger confidently predicted that genes would turn out to be molecules. But was it really absolutely necessary, he asked modestly, to invoke the foundations of quantum mechanics (his own invention) to come to this conclusion? Could not a nineteenth-century scientist, equipped with Mendel's laws and the empirical observation of molecules that persist for geological periods of time, have come to the same conclusion? Perhaps, Schrödinger admitted; but because nineteenth-century physicists did not understand what kept molecules together, their answer would have been '... of limited value as long as the enigmatic biological stability is traced back only to an equally enigmatic chemical stability'.

Schrödinger's incisive philosophical point is that genes cannot be said to be understood until molecules are understood. Biologists and chemists, who are permitted by their rules to ignore details of atomic structure that are irrelevant for their purposes, might disagree. Indeed, it is possible to explain most biological and chemical processes without having first mastered physics; but in the case of heredity, stricter accountability is required. If the aim is to find out whether life can be understood in materialistic terms, the chain of reasoning must go beyond the level of chemistry. If you want to plumb the secret of life, you'd better be prepared to get all the way to the bottom of things.

And to understand molecules, Schrödinger insists, you *do* need quantum mechanics. Why, for example, does a water molecule, which consists of an oxygen atom with two hydrogens stuck to it like Mickey Mouse ears, and which is constantly bombarded by a barrage of fellow molecules smashing into it at suicidal speeds, stick together? The answer relies on a variant of Planck's quantum hypothesis. The energy contained in the water molecule is not continuous, but divided into discrete levels like an irregular staircase. Each succeeding step corresponds to a higher state of excitation of the molecule into rotational and vibrational motion. The secret of the stability of water is that an incoming punch that delivers a quantum of energy below the lowest rung of the molecule's energy staircase has no effect whatsoever. A molecule doesn't disintegrate crumb by crumb like a biscuit; destruction is an all-or-nothing proposition, and most of the time it's nothing. This protective mechanism, revealed by the quantum hypothesis, was the missing link that rendered molecules comprehensible.

In summary, Schrödinger argued that genes were both small and stable, that stability could not be guaranteed to small atomic systems by classical mechanics, that molecules were also small and stable, that hereditary information must therefore be encoded in molecules, and finally that quantum mechanics furnished a complete explanation of why molecules persist. In this view, sudden, jump-like changes in molecular configuration allowed by quantum mechanics correspond perfectly to the sudden jump-like genetic changes observed as mutations.

The whole picture is remarkably coherent, but Schrödinger does not pursue it beyond this point. Physics has gone as far as it can in solving the riddle of life, he writes, and it is time to return the problem to the biologists:

> [The] molecular model, in its complete generality, seems to contain no hint as to how the hereditary substance works. Indeed, I do not expect that any detailed information on this question is likely to come from physics in the near future. The advance is proceeding and will, I am sure, continue to do so, from biochemistry under the guidance of physiology and genetics.

Only a decade later the zoologist-geneticist James D. Watson, working with the ex-physicist Francis Crick, identified the hereditary substance as DNA, and discovered the structure of its molecule to be a double helix. It was no accident that their most feared competitor was the two-time Nobel laureate Linus Pauling, who had written a textbook on quantum mechanics and contributed more than any other chemist to the understanding of what kept molecules together.

By now biologists, aided by a fundamental theory of matter as well as the diagnostic instruments furnished by physics, and the vast lore of particular properties of molecules accumulated by chemistry, knew that hereditary information was carried by chromosomes, and that chromosomes, in turn, consisted largely of DNA. But how? Human chromosomes are huge on the atomic scale, each set containing billions of the bases that make up their DNA. How do you build the command centre for human growth and development from that myriad of basic units? What is the architecture of heredity?

To solve this remaining mystery, nature obligingly offers another one of its clues, and *mirabile dictu* it is the same clue all over again on a finer scale of size. Just as chromosomes turned out to be linear arrays of genes, they are also linear arrays on the molecular level.

A single, immensely long double-stranded fibre of DNA extends unbroken through the entire length of a chromosome! If stretched and hung out to dry, it would be as long as your hand. To be sure, that long fibre has to be curled up in a tight, three-dimensional wad in order to fit into the confines of a chromosome, but it is nevertheless linear. That simple structure, like the lining-up of letters to form prose, underlies the modern, sophisticated methods of molecular biology that have led to genetic engineering and the charting of the human genome. Information about the positions of DNA's components was discovered in the same way as, on a larger scale, the structure of chromosomes was unravelled. Crossing, recombining, cutting, splicing and mutating all played similar roles again.

The simple, linear organization of chromosomes and DNA molecules made possible the progress from Mendel's laws to the human genome in a mere century. In contrast, other human information-processing systems, such as the visual perception apparatus and the brain, with their two- or three-dimensional structures, will be incomparably more difficult to unravel.

In *What is Life?* Schrödinger got it right: hereditary information is encoded in molecules; but in 1943 that notion was still so novel that he subtly pulled his punch. After quickly reviewing the requisite biology, and before embarking on his masterful argument about sizes and lifetimes of genes, he inserts a little demurrer:

> Of course, the scheme of the hereditary mechanism, as drawn up here, is still rather empty and colorless, even slightly naive. For we have not said what exactly we understand by a property. It seems neither adequate nor possible to dissect into discrete 'properties' the pattern of an organism which is essentially a unity, a 'whole'.

He just can't bring himself to subscribe to the reductionist claim that the entire incredibly complex structure of an eye, for example, can be quantized and expressed by means of a molecular code. To save his argument, he resorts to an ingenious qualification. In view of the fact that Mendel, Morgan, and all the other geneticists had studied mutations, or *changes* in well-defined properties, Schröd-inger restricts the function of genes to recording *differences* in properties, rather than the properties themselves: the switch from purple to white flowers, not the blossom itself, and the switch from blue to brown, not the structure of the whole eye.

Being of a philosophical turn of mind, Schrödinger promptly elevates this remark to a general principle: 'Difference of property, to my view, is really the fundamental concept rather than property itself, notwithstanding the apparent linguistic and logical con-tradiction of this statement. The differences of properties are actu-ally discrete ...' Later research demonstrated that Schrödinger's caution was excessive. His proposal to locate the genetic code in molecules proved to be so spectacularly successful, and provides for such massive amounts of information, that even the properties themselves can be accounted for, not just differences in properties.

The discovery of how the complex instructions for generating and maintaining life, the body of data collectively called 'her-editary information', are stored and transmitted, represents a triumph of twentieth-century science. It was made possible by a remarkable coincidence – the fact that nature mimics art in its linear display of data – and a proven method – the reductionist technique of investigating the parts in order to understand the whole. Since reductionism, in turn, is not without its critics, we pause at this point to make sure that our inquiry is heading in the right direction.

7 A Battle Among Giants

Reductionism and emergence

The protagonist of Italo Calvino's novel *Mr Palomar* thinks deeply about the world around him. Everything he sees, from the stars in the night sky and the cheeses in a Paris dairy shop to the naked bosom of a young sunbather, fills his mind with profound metaphysical musings. One day he contemplates his lawn. Looking closely at a patch of ground with its countless little blades of grass, he grows unsure even of the very concept of lawn: '... is "the lawn" what we see, or do we see one grass plus one grass plus one grass ...? There is no point in counting them, the number does not matter; what matters is grasping in one glance the individual little plants, one by one, in their individualities and differences. And not only seeing them: thinking them.' The challenge, he concludes, is to grasp the whole as the sum of its parts. The lawn, he believes, cannot be fully understood except as a collection of grasses.

Mr Palomar is seduced by the philosophical position called reductionism, and a seductive programme it certainly is. If you don't understand something, break it apart; reduce it to its components. Since they are simpler than the whole, you have a much better chance of understanding them; and when you have succeeded in doing that, put the whole thing back together again.

Will reductionism help us to understand information? Does the study of *bits*, or other basic units, reveal the real nature of the concept of information? Can we learn anything by following

Shannon, or do we have to proceed at once to the problem of meaning, which he sidestepped?

Physics again provides a clue, because nowhere has reductionism been more spectacularly successful than in the study of matter. Even as that science grew increasingly abstract, it continued to delve down to ever smaller and more primitive parts. Beginning with Democritus, who declared that the world was made of atoms and the void, reductionism has dominated that subject. Solids, liquids and gases are made of molecules, molecules are made of atoms, atoms of particles, particles of quarks and leptons, and those, in turn, of some primitive primordial stuff that may have the shape of a string. With the discovery of the double helix, living matter, too, has joined the reductionist chain of explanations. When modern scientists think of matter, they are in the same position as Mr Palomar: beneath the complex phenomena their senses perceive, they are always conscious of the underlying granular structure of the world.

Yet for all its explanatory power, reductionism has its limitations. To understand how a car works, it is pointless to analyse the chemical composition of its paint; to come to grips with human behaviour, you don't study atomic physics. In fact, for the logical connections between disciplines, the image of a chain is quite apt. Each link is affected by, and in turn affects, its nearest neighbours, but the longer the distance between two links, the less obvious is their relationship; and if the links are altogether too far apart, their relationship ceases to be apparent.

One author who failed to heed this caveat was the Roman poet Lucretius, whose long, lyrical poem *De Rerum Natura* (On the Nature of Things) served as an inspiring defence of atomism for over a millennium and a half. Countless puzzling phenomena, from the motion of fish through the continuous bulk of the water to the mysterious nature of the invisible wind, are treated with what today appears as amazing insight so long before the reality of atoms was commonly accepted; but then Lucretius overreached. Not content with painting a rational picture of the material world, he went on to explain, following his role model Democritus, how even the soul is made of atoms. His stubborn reductionism beyond all reasonable bounds did much to discredit the doctrine of atomism for centuries to come, and earned for him the fierce condemnation of the Church. Reductionism was regarded with as much affection as a scorpion in the baptismal font. It took a long time to repair the damage. Even as late as the beginning of the twentieth century it was still possible for serious scientists to question the reality of atoms, dismissing them as mere theoretical, though useful, fictions; but when the dam finally broke, it swept away all doubts. As far as matter is concerned – living matter no less than inanimate matter – the reductionist philosophy of Democritus and his prophet Lucretius has carried the day.

In the third quarter of the twentieth century, however, the debate over the role of reductionism in science flared up again. The event that triggered it seems to have been a 1974 article in the *Scientific American* by the Nobel laureate Steven Weinberg, the co-architect of the Standard Model of Elementary Particles. 'One of man's enduring hopes', he wrote, 'has been to find a few simple general laws that would explain why nature with all its seeming complexity and variety is the way it is. At the present moment the closest we can come to a unified view of nature is a description in terms of elementary particles and their mutual interactions.' This remark was subsequently cited by Ernst Mayr, one of the world's leading evolutionary biologists, as 'a horrible example of the way physicists think'. He called Weinberg 'an uncompromising reductionist', whereupon Weinberg took offence, and the battle was joined.

Weinberg is as stubborn as he is eloquent; in book after book and essay after essay he has been carrying on the fight for thirty years. When giants wrestle in the ring of academic debate, pearls of wisdom are sometimes flung out through the ropes. Perhaps the most useful contribution Weinberg made to the argument is a clarification of the terms of the dispute. He distinguishes two very different types of reductionism, which he calls 'petty' and 'grand'. (He borrowed the terms from criminal law, but they also apply to the two kinds of epileptic fit called *petit* and *grand mal*.) Petty reductionism, according to Weinberg, is 'the doctrine that things behave the way they do because of the properties of their constituents'. It is the reductionism of Mr Palomar, and of the first half of Lucretius's poem; but for many questions, such as the nature of the soul and the cause of human behaviour, it doesn't have much value. Being applicable only to a small set of questions, it isn't, in the end, such a big deal. Giants shouldn't really get into a fit over it.

Grand reductionism is much more interesting. It is the 'view that all of nature is the way it is (with certain qualifications about initial conditions and historical accidents) because of simple universal laws, to which all other scientific laws may in some sense be reduced'. In other words, not *things*, such as Mr Palomar's lawn, can be reduced to simpler components, but *laws*, the way Galileo's law of falling can be reduced to Newton's universal law of gravitation. Weinberg uses this conviction as a starting point for the defence of his own discipline of elementary particle physics. In his battle with Weinberg, the nonagenarian Mayr was eventually relieved as principal adversary by a physicist, fellow Nobel laureate Philip Anderson of Princeton University. As I write this, in July 2002, the latest instalment of the debate has just appeared in the journal *Physics Today* in the form of Anderson's book review of Weinberg's latest collection of essays.

Anderson, as might be expected, is not a particle physicist, but an expert on matter in the condensed state: solids and liquids in which billions of atoms jostle each other in a spectacle of daunting complexity. Anderson does not agree with the simple reductionism of Mr Palomar – the lawn, he would claim, is much more than the sum of its grasses. Ecologists would agree. For Anderson, it is the principle of emergence that holds the key, and, as a helpful caricature of his own point of view, Anderson coined a nifty aphorism worthy of John Wheeler: 'More is different.'

Anderson, even as he himself is engaged in the search for universal principles, claims that in complex systems new and totally unexpected laws may emerge. Consider, for example, a fair coin. There is absolutely nothing that can be said about whether it comes up heads or tails after a fair toss. We understand the experiment and we understand that its outcome is unpredictable. End of story. But for a hundred tosses a surprising new rule begins to show up: half of the results, with increasing accuracy as the number of throws goes up, are heads. A new law, a law of large numbers, has emerged. More is different.

A related but more physical example of emergence is furnished by the concept of temperature, on which the science of heat is based. For a single molecule, or for a small handful, the term 'temperature' doesn't even have a meaning: it can be defined only when particles swarm in prodigious numbers. The laws of thermodynamics are some of the most beautiful and general in all of science, but it is fair to say that they have not yet been deduced satisfactorily from the laws that govern individual particles. Of course, Weinberg knows this very well, but he has faith that thermodynamics is 'just' statistical mechanics, and that statistical mechanics – Anderson's discipline – is 'nothing but' particle physics combined with the mathematical laws of statistics.

Yet for Anderson, as he points out disarmingly, the world we perceive, the world of thermometers and automobiles, and of human behaviour and souls, does not resemble Weinberg's world of quarks and leptons in any way. There must, therefore, be many layers of emergence of novel properties of these complex, macroscopic things that are not explained by the Standard Model of particles. This idea – the emergence of laws – he calls 'The God Principle', in parody of the title *The God Particle* of a book about the Standard Model.

A compromise between the positions of the battling physicists is offered by another giant of science, the Harvard biologist Edward O. Wilson. Wilson is a latter-day Lucretius, as eloquent in prose as his Roman counterpart was in hexameter, and just as staunchly materialistic. Both display an almost religious zeal in trying to convert the world to their radical point of view. Wilson's book *Consilience*, a celebration of the fundamental unity of all knowledge, is a twenty-first century *De Rerum Natura*. However, while Wilson regards reductionism as 'the cutting edge of science', and even 'the primary and essential activity of science', reductionism is only a means to a higher end. Simplicity is not the highest goal; understanding the complexity of the world is. To make his point, Wilson proclaims: 'The love of complexity without reductionism makes art; the love of complexity with reductionism makes science.' (The form of this epigram ironically mirrors the rhythm of a famous Einstein quote that would leave Wilson cold: 'Science without religion is lame, religion without science is blind.')

Then Wilson explains his world view, which could pave the way for a reconciliation between Weinberg and Anderson. First, he connects petty with grand reductionism: 'Behind the mere smashing of aggregates into smaller pieces lies a deeper agenda that also takes the name of reductionism: to fold the laws and principles of each level of organization into those at more general, hence more fundamental levels.' But although the Holy Grail remains precisely as Weinberg claimed back in 1974, like Anderson, Wilson believes that this ambitious goal may be unattainable. At each level of complex organization, especially at the living cell and above, there are laws and principles that have not yet been derived from more general laws, and may never be. The prospect of failure doesn't faze Wilson, however. On the contrary, he confesses that the gaps in the explanatory chain of reasoning, which is supposed to connect quarks, lawns and the human soul, provoke his curiosity: 'The challenge and crackling of thin ice is what gives science its metaphysical excitement.'

Applied to information, petty reductionism requires first of all the identification of the smallest units or components. In the computer age there is no doubt that bits are prime candidates for serving as atoms of information, though simply counting them, as Mr Palomar realized, may be pointless. But bits are not the only option. In contexts other than computing and communication, different measures of information have been suggested and tried. Economists, as expected, measure information in dollars. In the field of Pattern Recognition, what matters is accuracy, so information is measured in hits – the number of *correct* identifications of a message bit. The most unfamiliar alternative to bits is required by quantum mechanics, where a new unit of information, the qubit, was introduced a decade ago. In the second half of this book I will deal with these strange little beasts.

All this, though, is in the service of a deeper agenda, the grand reductionist search for the fundamental principles of nature. In the last chapter I will describe a candidate for just such a law: Anton Zeilinger's conception of the way the classical, macroscopic world we know emerges from its underlying quantum mechanical substrate. Zeilinger's principle, couched in the language of information, can be cast as a really big question: IT FROM QUBIT?

8   The Oracle of Copenhagen

Science is about information

If Raphael were commissioned to paint a *School of Twentieth Century Physics*, in the manner of his monumental Vatican fresco *The School of Athens*, the central figures dominating the scene would be Albert Einstein and Niels Bohr in place of Plato and Aristotle. The theories with which their names are associated – relativity and quantum mechanics, respectively – form the twin supporting pillars of modern physics. The current quest for a synthesis of general relativity with quantum mechanics is reminiscent of the Renaissance search for a reconciliation of the two opposing camps of Greek philosophy, which inspired Raphael's painting; and whether or not it succeeds, which it conceivably could sometime in the coming decades, the clash between the views of Einstein and Bohr on the nature of physical reality will continue to reverberate.

In a fitting coincidence the two towering figures of physics received their laurels simultaneously: the 1921 Nobel Prize for physics was delayed for a year, and then awarded to Einstein in the same ceremony at which Bohr received the 1922 prize; but while their careers and contributions to science are inextricably intertwined, the personalities of the two men show remarkable contrasts. Einstein, though idolized by his colleagues and lionized by the public, always remained a loner. He had few collaborators and no students to carry on the research programmes he initiated. Bohr, on the other hand, established the Institute for Theoretical Physics at the University of Copenhagen, where two generations of physicists were to develop the quantum theory of the atom and, later, the foundations of nuclear physics. The influence of Bohr's school is commemorated in the name physicists attach to the predominant interpretation of quantum mechanics, the so-called 'Copenhagen interpretation', a compromise hammered out under Bohr's leadership by the theory's builders, all of whom passed through his institute.

The differences between Bohr and Einstein run far deeper than this, though. Indeed, they even extend to those attributes most closely associated with their names in the public mind. For as in Raphael's fresco, where Euclid is identified by the compass he wields and Pythagoras by the right triangle he ponders, the attribute universally associated with Einstein is the formula $E = mc^2$, while Bohr's name is tied to that ubiquitous icon of the atom, three overlapping ovals, representing electron orbits, around a central dot, which stands for the nucleus. Whether it is embossed on the now-obsolete Greek ten-drachma coin, printed on the letterhead of a small high-tech company, or monumentally rendered in aluminium on the wall of my building at the College of William and Mary, the visual representation of Bohr's planetary atomic model is unmistakable.

The contrast between these two popular scientific symbols could hardly be more pronounced. Einstein's famous equation exemplifies the lapidary succinctness of his pronouncements, and the equivalence of energy and mass that it expresses remains a correct and immutable cornerstone of the theory of relativity. The picture evoked by the atomic logo, on the other hand, is utterly wrong. Even though it lost its scientific relevance more than eighty years ago, and was repudiated by Bohr himself shortly after he invented it, it remains as popular as ever. The circumstances of Bohr's adoption and rejection of his atomic model illustrate his philosophical quarrel with Einstein about the nature of reality.

In 1913 the young Bohr had learned from Ernest Rutherford that the atom was not the solid pebble that everyone from Democritus on down had imagined it to be, but was, in fact, a hollow, wispy thing, whose weight resided almost entirely in a minute, hard nucleus at the centre of a large, insubstantial cloud of electricity. In an effort to explain this structure, Bohr proposed that a hydrogen atom's electron was a little moon orbiting the nucleus under the influence of an electrical attraction – a bold, if not preposterous, suggestion.

Bohr's model, unlikely as it may seem, made accurate quantitative predictions and launched the new discipline of atomic physics; but it ran into trouble as soon as it was born. Not only were its fundamental assumptions inconsistent with what was known at the time about electricity and magnetism, but its early successes with hydrogen and a few similar atoms could not be duplicated for more complex systems. Worse, even for hydrogen the fundamental picture turned out to be badly wrong. The hydrogen atom is not a flat, pancake-like affair, as the planar orbit of its electron would suggest, but a smooth, round ball. Furthermore, its negative electrical charge is not distributed along the electron's orbit in a sort of doughnut configuration, but is instead concentrated in the centre of the nucleus, just where the doughnut has a hole. By 1919, six years after its publication, Bohr was forced by such inconsistencies to reject his own model.

For the next six years the field of atomic physics floundered in disarray. The only clear, plausible model had been abandoned, and no replacement was in sight. Frustration ruled the day.

When quantum mechanics burst upon the scene in 1925, relief swept over the community of physicists. The new theory relegated the Bohr atom to the attic of discarded models to join the flat Earth and the geocentric solar system, and opened the prospect of reducing atomic physics to the solution of abstract, well-defined, and increasingly complex calculations. At the same time, however, the solid ground of intuitive understanding dissolved. The Bohr model, with its visual appeal and visible analogue in the form of the Moon overhead, gave way to a quagmire of fresh doubts and confusion.

The debate about the meaning of quantum mechanics, which started with the birth of the theory, has not abated in the intervening seventy-five years and shows no sign of drawing to a close. While everyone agrees on the theory's technical content and unblemished record of practical utility, opinions about its meaning cover a wide spectrum. The extremities of this spectrum are delimited by the answers to a metaphysical question: *Underneath the shifting appearances of the world as perceived by our unreliable senses, is there, or is there not, a bedrock of objective reality?*

Einstein felt strongly that there was, and further that it was the business of physics to discover that ultimate reality. Since quantum mechanics, with its intrinsic randomness and apparent observer-dependence, seems to deny us access to such a level of certainty, he believed that the theory must be incomplete, and would eventually have to yield to a more fundamental understanding. In spite of the unbroken string of successes of quantum mechanics, there are some physicists today who sympathize with this position, and put their hope in the future.

A more pragmatic attitude, which enjoys a large following among today's practitioners, was that of Richard Feynman. Although he demonstrated his superior understanding of quantum mechanics by creating a completely new and characteristically unconventional mathematical formalism for dealing with it, he was unable to shed new light on its interpretation. Feynman did not indulge in polemics about meaning, but did confess to private misgivings: 'It has not yet become obvious to me that there is no real problem. I cannot define the real problem, but I'm not sure there is no real problem.' But by his actions, Feynman implied that we should put such worries aside and get on with calculating things we can measure.

Niels Bohr could not shrug off his doubts as glibly as Feynman. After his clear, palpable image of the atom had gone up in flames, he was forced to watch a very different reality begin to emerge under his own leadership. How could he reconcile this new world view of quantum mechanics with his own yearning for certainty and clarity?

The Copenhagen interpretation answers the metaphysical question about the nature of the world with a resounding *no*: objective reality is an illusion we construct for our own comfort. The best we can do, Bohr came to believe, is to create a coherent model of the world that reproduces its measured properties without claiming to describe reliably what actually *is*. In a letter to a colleague he explained what he believed to be the mandate of science: 'Our task is not to penetrate into the essence of things, the meaning of which we don't know anyway, but rather to develop concepts which allow us to talk in a productive way about phenomena in nature.' Elsewhere he expressed the same thought more forcefully: 'There is no quantum world. There is only an abstract quantum physical description. It is wrong to think that the task of physics is to find out how nature is. Physics concerns what we can say about nature.'

To illustrate what Bohr meant, consider the hydrogen atom. The question is no longer: 'How is the atom really constructed?' but: 'What is the probability of finding the electron here, or there?' The ultimate reality, according to Bohr, is not the thing in itself, but the sum total of our information, quantified in terms of probabilities, about the thing. The wave function, which encodes this information, therefore has more claim to 'reality' than the image of an electron circling around the nucleus like a planet.

In fancy words, Bohr believed that physics is not about ontology, the science of essences, but about epistemology, the study of how we know what we know, and of the limitations to our knowledge. Epistemology, in turn, has always been concerned with the flow of information. Where Aristotle explained how the form of a stone enters the human soul, today we wonder about the process of extracting information from an atom and transferring it into the brain. Bohr's philosophical position, which he passed on to John Wheeler, who in turn inspired Anton Zeilinger, can be generalized to include not just physics, but all of science: *Science is about information.*

Since Einstein's world view – that science is about ultimate reality – is more reassuring, and much easier to put into words, Bohr has often been criticized as a modern-day Sybil, sitting in his temple in Copenhagen issuing oracular pronouncements in muddled phrases. Thus the Nobel laureate P. A. M. Dirac once said: 'While I was very much impressed by [him], his arguments were mainly of a qualitative nature, and I was not able to really pinpoint the facts behind them.' John Stuart Bell, who two years after Bohr's death initiated the modern era of experimental investigation of quantum mechanics, called Bohr an 'obscurantist'. Bohr formalized his difference with Einstein by proposing, only half in jest, an Uncertainty Principle of Scientific Knowledge: that *Klarheit* and *Wahrheit*, clarity and truth, are complementary. Increasing one automatically decreases the other. From the fate of his atomic model Bohr had learned the painful lesson that clarity can mislead, and that the truth may hide under a fog of confusion. How ironic, then, that it is the model that has become his most lasting legacy!

In the end, the most accurate assessment of Bohr's role in modern physics was left to his friend, intellectual sparring partner and only peer, Albert Einstein: 'He utters his opinions like one perpetually groping and never like one who believes he is in possession of the definite truth.' Einstein knew that, while science must strive for certainty, it thrives on uncertainty, ambiguity and doubt. Attaining its goal would kill it.

So we grope our way toward an understanding of the nature of classical information.

Classical Information

How probability measures information

The theory of probability became a serious science in the seventeenth century, but its roots are buried in the ancient games of dice and cards. Long before mathematicians refined their tools sufficiently to tackle the subject, gamblers knew from experience how to use partial information as a basis for estimating the odds on their bets. In this way they intuitively assigned quantitative values to the information they had, or lacked, about whatever they were betting on. In view of this connection, it is perhaps not surprising that the most dramatic demonstration of the subtle way in which betting odds reflect information occurred on a TV game show – not directly, as it happened, but via a magazine article that described it.

Marilyn vos Savant writes a popular column on mathematical games and logical brain-teasers for *Parade* magazine. That she is also a self-proclaimed genius with an alleged IQ of 228 doesn't detract appreciably from her considerable following. On 9 September 1990 she posed a puzzle: Monty Hall, host of the TV show *Let's Make a Deal*, shows a lucky contestant three cubicles, numbered from one to three and closed off by curtains – one of which conceals a brand-new car. If he can correctly guess its location, he wins it. At random the contestant picks cubicle number 1. The host, *who knows where the car really is*, now opens cubicle number 2, showing him that it is empty, and asks generously: 'Do you want to stick with your first guess, cubicle number 1, or switch to cubicle number 3?'

Gut reaction suggests that since the car could be located behind either of the two curtains, his probability of success is 50 per cent either way, so it doesn't matter whether he sticks or switches. But that answer is wrong. Imagine a savvy audience mouthing advice: 'Switch! Switch!' By switching his choice to number 3 the contestant would, in fact, increase his chances of winning from 1 in 3 to 2 in 3, as Ms vos Savant proceeded to explain in her column.

The article touched a nerve. She received bags of mail and was ridiculed by university professors and other certified 'experts' on probability, some of whom pilloried her as a mathematical ignoramus and chided her for enriching herself by misleading the public. Even long after she had amplified her argument in two subsequent columns, the 'Monty Hall Problem' remained a cult classic hotly debated in the media and the technical mathematical literature, as well as in bars, classrooms and university hallways. Inevitably it found a home on the World Wide Web, where it still provokes impassioned discussions and inspires the creation of elaborate simulations for empirical investigation.

The popularity of the problem is not difficult to understand. Just as mathematicians and physicists value an argument that begins with the obvious and proceeds, by rigorous logic, to prove the incredible, amateurs are drawn to simple problems with counter-intuitive solutions. Such conundrums seem to make contact with something beyond our feeble human comprehension, something larger than ourselves, an intimation of the absolute. They allow a rare glimpse of a Platonic Mathematical Truth, a true 'Form', somewhere above the human sphere.

The solution of the Monty Hall problem hinges on the concept of information, and more specifically, on the relationship between added information and probability. Imagine that you are the contestant and that your first pick, as stipulated, is cubicle number 1. At this stage there are three equally likely possibilities (and your chances of success are, accordingly, 1 in 3):

A. The car is in cubicle number 1.

B. The car is in cubicle number 2.

C. The car is in cubicle number 3.

However, this soon changes, for as soon as the host opens a curtain, adding information to the statement of the problem, the odds will change. In the first case, whichever curtain Monty opens, you are clearly better off sticking with your first choice. But in case B, Monty will open the third curtain, showing an empty cubicle, and you would lose if you stuck. Similarly, in case C Monty opens the second curtain, and again you would lose if you stuck. So in two cases out of the three, sticking with your original choice will lose you a car.

This explanation is, however, but a thumbnail sketch of the full story. The issues are so subtle that even after one has accepted the solution, understanding blinks into and out of focus in the most disconcerting way. A more intuitive demonstration of the effect of adding information makes use of the mathematical trick of exaggerating the assumptions to highlight them. Imagine that there are not three, but a thousand curtains, and one car. Initially you pick, say, number 815 with a resigned shrug – realizing that your chances of success are one in a thousand. The host (who knows precisely where the car is) now opens 998 empty cubicles. Not the one you have already picked, of course, and not cubicle number 137. Now he asks politely: 'Do you want to stick with your first guess, curtain number 815, or switch to curtain number 137?'

What do you think? Doesn't that lone closed curtain picked by the host, out of a long row of a thousand, fairly scream at you? Do you really feel it would be wise to stick to your first random choice? And if not in this case, why not in the original puzzle? The difference between the two versions is not in kind, but in degree. In the original puzzle you only doubled your chances by switching, whereas here you jump all the way from one in a thousand to virtual certainty.

A poignant footnote to the saga of the Monty Hall problem concerns the late eccentric genius Paul Erdós, one of the great mathematicians of the twentieth century. Although Erdós was not a pauper, he chose to live a peripatetic life relying on the kindness of fellow mathematicians. When he learned of the Monty Hall problem, he couldn't get over his immediate reaction – that the chances for curtains number 1 and number 3 must be fifty-fifty – and staunchly refused to yield to the explanations and computer simulations of his hosts of the moment. 'You haven't shown me *why* I should switch,' he kept lamenting. Valuing friendship over pride, they gave up and dropped the subject.

The point of this astonishing story is not that geniuses can err. 'Even excellent Homer nods,' is the way Horace put it. The point, rather, is twofold: first, that probability is an infernally tricky concept even in the simplest circumstances (such as a silly game show), and secondly, that probability is a way to quantify information.

Since the Monty Hall problem is so elusive, it is worth while to consider a simpler illustration of the same point. A father announces: 'I have two children, born three years apart, one of whom is a boy.' What is the probability that the other child is also a boy? One half, you reply, since the other one could be a boy or a girl. Wrong. Careless thinking, which deprived you of that new car, is misleading you again. The correct answer is one third. The man's two children could be a boy followed by a boy, a boy followed by a girl, or a girl followed by a boy; in short, BB, BG, or GB. Of these three possibilities, only the first contributes to the answer. (I vividly recall a night on a hiking trip into Virginia's Blue Ridge Mountains when, seated at the camp fire, I proved this claim to my children by tossing two pennies a hundred times.)

Suppose, furthermore, that the father then adds the following extra qualification: 'The taller of my two children is a boy.' Now the possibilities are reduced to {tall boy and short boy} and {tall boy and short girl}, and the probability that his other child is also a boy changes from 1 in 3 to 1 in 2. In contrast, suppose that, with the same two statements from the father, the question was amended to: What is the probability that the other child is a girl? In this case, the second response would remain at 1 in 2 but the initial answer would now be 2 in 3. Thus, astonishingly, the addition of information on one child's height has brought about a respective rise and fall in probability – despite its apparent irrelevance to the question of gender.

These examples demonstrate that information affects the computation of probabilities, sometimes in unexpected ways. The reverse is also true: probability is related to the very definition of information. The first person to make this connection in the context of science was the nineteenth-century physicist Ludwig Boltzmann, who suggested that probability arguments must be used when dealing with atoms precisely because their unimaginably large number leaves us with vast amounts of missing information. The connection Boltzmann only hinted at became manifest in the twentieth century when the question turned from 'What is information?' to 'How do you measure information?' and, more precisely, 'How do you measure the value of information?'

We feel intuitively that the import of information contained in a message is determined by what we expected in the first place; in fact, all of us evaluate information by estimating probabilities every day. When, for example, my daughter announces that there is no school today, and I recall that it is Sunday, I know that the probability of her announcement being true is 100 per cent, so I ignore her. If it is an ordinary weekday, the probability that she is telling the truth is practically nil, and I don't react either; but if I notice that it is snowing, the probability that she might actually be right rises to somewhere around 50 per cent and I take her very seriously. The value of the information she offers me depends on what I know, and what I know is expressed in the form of the probabilities that certain propositions are true.

If its relationship with probability is to throw light on the concept of information, it behoves us to clarify first what we mean by the word 'probability'. Most scientific and technical discussions define probability in terms of frequency – a simple and eminently reasonable approach. What is the probability of throwing heads in a hundred tosses of a penny? Divide the frequency of occurrence (50) by the number of cases (100) and you get the probability 1 in 2. What is the probability of throwing a seven with the toss of two

are ever to determine the probability of any particular proposition.' A small handful of scientists, mathematicians and philosophers have been wrestling with this challenge for more than two centuries, achieving considerable success but, apart from a few devotees, a remarkably muted response from the public or the broader scientific establishment. Physicists, for example, almost invariably continue to define probability in terms of frequencies. In order to get around the objection that when they analyse a specific experiment, they are, after all, talking about *one* vessel of gas, about *one* reading of its temperature and pressure, and about *one* measurement of its volume, they invent a huge, imaginary collection of identical copies of the container, call it an 'ensemble' and discuss its properties as a whole. In this way they are able to apply frequency arguments to singular events. The subterfuge works in most cases, because one can at least imagine repeating, or copying, a laboratory experiment; but for the economist studying the national budget this trick is more dubious.

The solution to Keynes's problem lies in the deployment of a numerical measure that expresses 'the degree of rational belief in the correctness of a specific inference from a given hypothesis'. This number, the probability, is symbolized by 'p', and is defined without reference to frequency, counting or statistics. It is not assigned arbitrarily, but by rational argumentation, though in many cases it is no more than an informed guess. It is the number called 'probability' in the four examples of unique events I listed above. For convenience I shall call this the 'rational definition' in order to distinguish it from the traditional, frequency-based 'statistical definition' – with no implication that the latter is irrational. Rational probabilities are what bettors use – they think about the outcome of an event, adduce all the information they can possibly muster, and assign a number to the degree of their belief. The value of p varies from zero, meaning that the event will certainly not occur, to one, corresponding to certainty that it will. British bookmakers are prepared to give you the p value for events as arcane as the probability that Tiger Woods loses an arm, a calamity that most definitely cannot be captured statistically.

The rational approach to probability supersedes its statistically based counterpart for two main reasons. First, by simply taking into its corpus those probabilities that can legitimately be defined by counting frequencies (thus making it more general), and secondly, by achieving practical successes in special cases where traditional methods run into conceptual or practical difficulties. For example, the branch of physics called statistical mechanics, which deals with systems of many particles, has been put on a firmer logical basis by being cast entirely in terms of rationally defined probabilities. In this context their determinations turn out to be 'not as simple as those for rolling dice, nor as complex as those for predicting the weather'.

The most celebrated consequence of the rational method of defining probabilities is a theorem named after its creator Thomas Bayes (1702–61), an English Nonconformist minister who was also an excellent mathematician. (In his honour, a rationally defined probability is often called a 'Bayesian probability'.) His work on probability was limited to a single essay published two years after his death and his theorem lay forgotten for many years after, although occasionally some famous scholar would dust it off in order to refute it. It is not part of 'school mathematics', the body of work routinely taught in schools and universities, and many scientists have never heard of it. Only in the recent past has it surfaced again, but its usefulness in economics, computer programming, medical research, and bioinformatics may yet elevate it to a much more visible perch.

Bayes' theorem answers the following question: Suppose you know, or assume you know, the probability that a certain conclusion follows from an initial hypothesis. Suppose further that a new bit of *information* is obtained, and added to the hypothesis. How do you then compute the updated probability that the conclusion is true, based on the combination of the old hypothesis with the new information? The value of the simple formula Bayes derived in answer to this question lies in its rigour. Although some of its ingredients (rationally defined probabilities) may involve guesswork, the way they fit together does not: the theorem is as indisputable as 2 + 2 = 4. The most important practical implication of Bayes' theorem is that it obliges you to compute, or at least estimate, an initial probability, called the *prior* probability. Without this step, the theorem makes no sense. It doesn't ask: how likely is this? – only: how is the likelihood changed by the new information?

Bayes' answer, as contained in his celebrated theorem, is a bit cumbersome to express, but don't worry. Immediately after its statement I will give a dramatic example to illustrate the idea.

In essence, the theorem states that the new probability p' (sometimes called *posterior* probability) for belief in the conclusion C, based on added information I, is equal to the *prior* probability p, multiplied by a factor. If that factor is greater than one, the new information increases rational belief in the conclusion; if it is less than one, rational belief is diminished. The factor, in turn, is the ratio of two other probabilities and is written $p(C{\rightarrow}I)$ divided by $p(I)$. The number $p(C{\rightarrow}I)$ is the key ingredient in the calculation and represents a kind of inverse of the desired answer: it is the probability that *if the conclusion C were assumed*, the information I would follow. The final ingredient $p(I)$ is the probability that the information I is true, regardless of any other assumptions.

Let's pursue one example, taken from medicine, to its startling conclusion. Suppose that a certain cancer has an incidence of one in a hundred, so the prior probability that you have it, called p, is 1 per cent. Suppose further that a new test for this cancer has been developed, and boasts an accuracy of 99 per cent. If you are unfortunate enough to test positive, an alarming piece of information we shall call I, what is the (posterior) probability p' that you really do have cancer? How much, in other words, should you worry? The unexpected answer follows from Bayes' theorem.

The number labelled $p(C{\rightarrow}I)$ is the likelihood that *if* you really had cancer, the test result would be positive. Since the test is so good, that probability is very high, so we set $p(C{\rightarrow}I)$ approximately equal to one. The last element, $p(I)$, is trickier. Out of a hundred people, one is expected to have cancer, and would probably test positive, while approximately one other healthy person, called a 'false positive', would also test positive because the test errs once in a hundred cases. So on average two people in a hundred test positive, regardless of whether they have cancer or not. This means that p(I), the probability that you test positive, regardless of other assumptions, is 2 per cent. The multiplying factor in Bayes' theorem is therefore approximately $p(C{\rightarrow}I)/p(I) = 1/(2\%)$. The probability that you are ill, calculated as the prior probability times the Bayes factor, is px $[p(C{\rightarrow}I)/p(I)] = 1\%x [1/(2\%)] = (1\%)/(2\%) = 1/2$. The chance that you are perfectly OK turns out to be 50 per cent, even though the test, which is supposed to be virtually infallible, was positive. Since this example is not a game, but could conceivably touch the lives of your family and friends, I have translated it back into the more familiar language of frequencies for a sample population of 10,000 in an endnote. The first column of its little table tells the amazing story.

Bayes' theorem is the perfect instrument for solving the Monty Hall problem. The prior probability of finding the car is easily guessed: it is 1 in 3 for each cubicle. What happens to the probabilities for cubicle number 1 and 3 when the host adds new information by opening cubicle number 2? The solution to this mathematical problem must explicitly take into account the effect of Monty's actions on each choice you can make. Using Bayes' theorem it takes just a few lines of calculation to confirm Marilyn vos Savant's answer. What this proof lacks in simplicity it makes up for in rigour and persuasiveness. I wonder whether it would have convinced Paul Erdós.

The Bayesian method is slowly gaining adherents in industry, especially in the field of medical and pharmaceutical research. Its distinctive feature is that it mixes new information – often in the form of statistical data – with a large body of prior experience that is not included in the new data. Bayesians believe that broad conclusions should not rest on limited statistical evidence, but should factor in a reasonably weighted estimate of prior probabilities, rationally defined. A recent example of the use of Bayesian methods was the analysis of a clot-buster – a drug used to improve the chances of surviving a heart attack. In 1992 a Scottish team of researchers had reported that one such drug had reduced the death rate of heart-attack victims by an astonishing 50 per cent. The sample of 300 patients was small, but the research seemed sound. Nevertheless, two British statisticians were sceptical, and decided to submit the result to a Bayesian analysis by comparing the rational expectation for the efficacy of clot-busters before and after the Scottish study.

Their contrarian conclusion – that the drug would reduce mortality by a mere 27 per cent, not the claimed 50 per cent, was published, and that seemed to end the matter without resolution. Then in May 2000, the *Journal of the American Medical Association* published a new review of a large number of clot-buster studies. It concluded that the drug actually reduced mortality by 17 per cent, much closer to the Bayesian result than the first enthusiastic endorsement. 'It's a real-life example,' exulted Robert A. J. Matthews, a physicist and Bayesian, about this story. 'People couldn't say, "Well, you just altered your prior probability [i.e. your guess] until it matched what we now know is the truth!" ' In other words, the Bayesian approach had yielded a far more accurate prediction – the ultimate triumph of a scientific theory.

The Bayesian approach allows human insight, subjective though it is, to be combined with statistical information, limited though it may be. It is not surprising that this blurring of the line between the methodologies of the sciences and the humanities has attracted passionate supporters as well as furious enemies on both sides of the cultural divide.

The conduct of science would be much improved if all its practitioners learned to use Bayes' theorem in the routine evaluation of their work. It is precisely suited to their needs, since science can never make absolute statements. The most that can be said is that its laws, principles, theories, conjectures and conclusions have certain prior probabilities of being right, ranging from zero to one. And every new experiment and new theory results in the acquisition of a new bit of information that alters those probabilities. If the change is large, the experiment is called crucial, or revolutionary. If, on the other hand, it is small, the experiment is correspondingly less important. Thus Bayes' theorem offers scientists a way to quantify the significance of their work, a step they may not always welcome; but by constantly reminding them to estimate prior probabilities, it forces them to be mindful of the contingent nature of their craft, and, more importantly, to state their hypotheses with sufficient clarity and explicitness to enable them to compute the four ingredients of Bayes' theorem. Physicists, who take pride in the quantitative nature of their world view, should do no less.

The scientific puzzle to which Bayesian thinking may make its most significant contribution is the interpretation of quantum mechanics. A fundamental problem of quantum theory concerns the meaning of the wave function, which by its nature is concerned with the prediction of probabilities. Accustomed to the frequency-definition of probability, most physicists therefore insist that wave functions apply only to large collections of particles, or frequently repeated experiments on a single particle, or at least a large number of imagined repetitions of the same experiment. On the other hand, the wave function makes definite correct predictions about experiments that involve only a single particle. It's almost as though the wave function guides the particle along its proper path. The clash between these two views has haunted the theory for three-quarters of a century. But if probability is redefined in Bayesian language, which allows probabilities to be assigned to singular events, the difficulty dissolves. If Wheeler's question – Why the quantum? – is to be answered in terms of information, then the posthumous essay by an obscure eighteenth-century cleric may turn out to play a crucial role.

The ubiquitous logarithm

Like astronomy and atomic physics, information technology must grapple with numbers too large for human comprehension. They can be brought down to a manageable level by the use of a mathematical curve called the natural logarithm. Since its invention in the seventeenth century, when it vastly increased the computational power of astronomers, the logarithm has become an indispensable practical tool in science, cropping up in a wide variety of different fields and hinting at formal connections where none would be suspected. Inasmuch as many of the things the logarithm describes impinge on our daily lives, we would do well to learn its lessons and to incorporate them into our thinking.

On graph paper, the logarithm and its inverse, the exponential, a smooth curve through the points of the geometric series, mirror each other across the diagonal line from the lower left to the upper right. Viewed in one direction, the exponential doubles, and then doubles again towards infinity; in the other, it shrinks and halves, approaching zero but never reaching it. The exponential's breakneck rise describes some of the characteristic phenomena of our time: the information explosion, the hydra-headed proliferation of AIDS viruses, the development of an embryo by cell division, the growth of money by compound interest, and the runaway chain reaction of fissionable material in an atomic bomb. At the same time, its diminishing tail characterizes both radioactive decay and the fading peal of a church bell.

Like the exponential, the logarithm always rises from left to

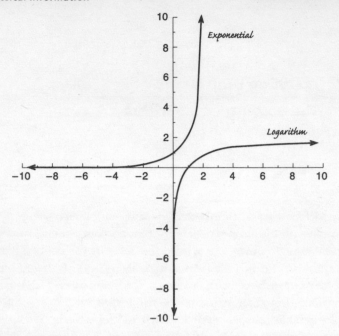

right. But whereas the exponential roars unchecked to infinity at an ever-increasing rate or slope, the rise of the logarithmic function is accompanied by a slope that gets continuously flatter; and whereas the exponential approaches the horizontal axis to the left of the origin like a landing glider that never quite touches down, the logarithm plunges precipitously through zero to negative infinity with increasing steepness, hugging the vertical axis ever more closely, though never quite reaching it.

At their extremities the exponential and logarithmic curves diverge dramatically, but near the origin they approach each other, running momentarily in parallel. Together they form the graceful outline of the thin waist of an hourglass.

The utility of the logarithm is aptly demonstrated by its ability to represent numerical excesses in comprehensible terms. To see how that works, consider a close cousin of the natural logarithm, called the logarithm to the base ten, or in its abbreviated form, the log. For multiples of ten, the log simply counts zeros, recording them as positive when they appear in the numerator and negative when they appear in the denominator. Thus the log of 1000 is 3 (written $1000 = 10^3$), the log of 1/100 is −2 (written $0.01 = 10^{-2}$), and the log of 1 is zero (written $1 = 10^0$). More generally, the log counts the digits of a given number, at least approximately. The rule is easy to remember: if you come across a number larger than one, written in ordinary decimal notation, you can find the approximate value of its log *by simply counting its digits* to the left of the decimal point.

A plot of the log has remarkable properties. On graph paper divided into one-centimetre squares, a point on the curve a mere eleven centimetres (or half a page) above the horizontal axis, lies 100 billion (10^{11}) centimetres to the right – reaching past the orbit of the Moon. Conversely, at a point eight centimetres below the origin the curve has moved to within 100 millionth (10^{-8}) of a centimetre, or an atom's diameter, from the vertical axis.

Scientists long ago adopted those properties with the 'powers of ten' notation. It is a marvellously convenient shorthand – one that prevents errors and saves space. Imagine calculating the properties of the early cosmos in longhand, writing out in full decimal notation the values of the Planck time (10^{-43} seconds) and the Planck length (about 10^{-35} metres) with their combined total of seventy-eight zeros – an impossible feat of patience and care.

Beyond this, the log further simplifies computations by converting arithmetic into counting digits. For example, 600 *times* 6000 equals 3,600,000 but three digits *plus* four digits equals seven digits. So if you are only interested in rough approximations, and a count of its digits is a good enough indicator of the magnitude of a number, then you never have to learn to multiply. Division is similarly simplified by becoming subtraction. This marvellous property of the log is responsible for the design of that ancient trademark of the engineer, the slide rule, which achieves multiplication and division of numbers by mechanically adding and subtracting their logs. Although the slide rule has now become obsolete, the log is still very much with us as an effective analytical tool for understanding the world, a way of looking at things that might be called 'logarithmic thinking', or more simply 'power thinking'.

Power thinking about the natural world began in the second century BC when the Greek astronomer Hipparchus divided the stars into six categories of brightness. The bright star Antares, Hipparchus declared, was arbitrarily classified as falling in the first magnitude of brightness. Polaris, visibly dimmer, but not by much, was deemed to be a second-magnitude star. And so on. (The modern scale of visual stellar magnitudes has been extended another thirty steps in one direction to include the Sun, and twenty-four in the opposite direction to capture the faintest object recorded by the Hubble space telescope, no brighter than a firefly at a distance equal to the diameter of the Earth.) Of course, Hipparchus had no way of determining brightness objectively, but his subjective classification turned out to be intrinsically logarithmic: as recorded by a detector that measures radiant energy, Antares is 2.5 times as bright as Polaris, which in turn is 2.5 times as bright as a third-magnitude star.

Hipparchus's original scale, however, illustrates an odd phenomenon: the human senses seem to perceive the world in a roughly logarithmic way. The eye, for example, cannot distinguish much more than six degrees of brightness; but the actual range of physical brightness covered by those six degrees is a factor of 2.5 × 2.5 × 2.5 × 2.5 × 2.5, or about 100. A scale of a hundred steps is too fine for human perception – Hipparchus's coarser intervals are about the smallest it can distinguish. The ear, too, perceives approximately logarithmically. The physical intensity of sound, in terms of energy carried through the air, varies by a factor of a trillion (10^{12}) from the barely audible to the threshold of pain; but because neither the ear nor the brain can cope with so immense a gamut, they convert the unimaginable multiplicative factors into a comprehensible additive scale. The ear, in other words, relays the physical intensity of the sound as logarithmic ratios of loudness. Thus a normal conversation may seem three times as loud as a whisper, whereas its measured intensity is actually 1,000 or 10^3 times greater. For this reason the decibel scale of loudness, invented by the telephone pioneer Alexander Graham Bell, is also logarithmic.

If sensory perceptions of brightness and loudness both depend logarithmically on physical stimuli, do they hint at some deeper, more fundamental law? For a century and a half that question has been at the frontier of psychophysics, a discipline founded by the nineteenth-century German physician and physicist Gustav Theodor Fechner of Leipzig, in which both of those disciplines overlap. Fechner had grappled for some time with the relation between stimulus and response, and while still in bed on the morning of 22 October 1850, so the story goes, he came up with a solution to the conundrum.

Fechner's law stipulates that the magnitude of a sensation – brightness, warmth, weight, electrical shock, any sensation at all – is proportional to the logarithm of the intensity of the stimulus, measured as a multiple of the smallest perceptible stimulus. (The latter condition, that one stimulus is measured as a multiple of another – for example, as three times the smallest perceptible one – is clever. That way the stimulus is characterized by a pure number – three, in our example – instead of a number endowed with units, like seven pounds, or five volts, or 98 degrees Fahrenheit. By purging his law of its dependence on specific units, Fechner was able to come up with a general rule that applies to stimuli of different kinds.) For more than a century Fechner's law was regarded as the supreme principle of psychophysics, as important to that discipline as Newton's second law is to physics.

But psychophysics is not simple physics. Beginning in the 1950s, serious departures from Fechner's law began to be reported, and today it is regarded more as a historical curiosity than as an ironclad rule. But even if Fechner's law is not accurate in every detail, it remains emblematic of a profound relation. If the intensity of the material world is plotted along the horizontal axis, and the response of the human mind is on the vertical, the relation between the two is represented by the logarithmic curve. Could this rule provide a clue to the relationship between the objective measure of information, and our subjective perception of it?

Logarithms can help make sense of the universe in another way, as well: they enable the inconceivably large and the infinitesimally small to register in our consciousness. Perhaps the most famous modern example of power thinking grew out of a children's book, *Cosmic View: The Universe in Forty Jumps*, written in 1957 by a Dutch schoolteacher named Kees Boeke. The book consisted of forty images of the world at consecutively larger scales, beginning with nuclear dimensions (10^{-15} metre) and ranging out to the edge of the visible universe (10^{25} metres). The designer Charles Eames, famous for his moulded plywood furniture, together with his wife and partner Ray, turned Boeke's book into an eight-minute film, which in turn inspired the wonderful 1982 book *Powers of Ten*, written by the Eameses and another husband-and-wife team, Philip and Phyllis Morrison.

What *Powers of Ten* did for space, another more recent book, *The Five Ages of the Universe*, does for time. It divides the history of the universe into five epochs, measured not in years, but in logarithms of years called cosmological decades. That unconventional reckoning of time has some notable implications. First and most remarkably, the Big Bang is not part of the history of the universe. Since the log of zero is too far down the vertical axis to be plotted on any finite graph, logarithmic reckoning comes as close as one desires to the moment of genesis – but it does not get there. From the scientific point of view, that limitation may be a good thing. Observation and analysis enable cosmologists and astrophysicists to approach the Big Bang backwards from the present, but the closer they get, the weirder the physics becomes. The times of $10^{-\infty}$ years and, for that matter, $10^{+\infty}$ years, too, both lie outside the range accessible to science.

The Nobel laureate Sheldon Glashow of Harvard University, whose work as an elementary particle physicist exposes him daily to numerical extremes, illustrates power thinking closer to home by showing how naturally it deals with a human life. Dividing life's development into nine stages (fertilized egg, free blastocyst, attached blastocyst, embryo, fetus, infant, child, teenager, adult), Glashow notes that on a normal, linear plot over, say, seventy years, the first six stages barely show up; but when the same stages are recorded logarithmically, they occupy roughly equal intervals on the vertical axis (except the first, which begins with the moment of conception at time $10^{-\infty}$ years). Thus the log properly reflects the universal cosmic and human truth that things take place faster at the beginning than near the end. As Glashow put it: 'A month is an eternity to a six-year-old child, but it passes in a twinkling to a pensioner.'

Frank Wilzcek, across the Charles River from Glashow at MIT, relies on the logarithm to reach across a different divide than that which separates the embryo from the pensioner. He points out that we can come to terms with the incomprehensibly tiny strength of gravity compared to atomic and nuclear forces – a mere one part in 10^{18} – by reference to the prevailing theory of quarks and leptons. Deeper understanding, he claims, takes the surprise out of the 'extravagance' of that tiny number, and makes it appear natural. The linchpin of his argument is the theoretical prediction that, as we descend the ladder of distance to the subnuclear realm, the strength of forces changes much more gently than expected and, in fact, follows a logarithmic curve. Thus power thinking can come in useful in solving one of the long-standing puzzles of modern physics.

In a more familiar context, the log also finds application in the Richter scale of earthquakes. There are only ten steps in this scale, but each one implies an increase in energy release by a factor of thirty-two. Accordingly, the ratio between a seismic catastrophe and a barely noticeable tremor, measured in units of energy, is a whopping 32 trillion. Since no one can imagine numbers ranging over tens of trillions, the Richter scale of earthquake magnitudes makes use of the log to create a comprehensible classification.

What is it about the log that makes it relevant to so many disciplines: from astronomy, acoustics, cosmology and particle physics to embryology and even geology? What could those disparate fields possibly have in common? The answer, of course, lies in the unifying role of mathematics. Many processes in nature and technology unfold linearly or arithmetically, like bricks added to a structure one by one. Time is like that, and spatial extension, and money, and the number of atoms that take part in some physical transformation. However, other processes change geometrically, exponentially or multiplicatively: probabilities and explosions, compound interest, populations and proliferating neural connections – wherever one parent engenders more than one offspring. The logarithm and its inverse, the exponential, are the simplest mathematical functions that mediate between those two kinds of process. The surprise would be if the log did *not* show up, at least as a first approximation, in unrelated niches of science and technology.

In the coming century, astronomers, economists and computer scientists will inundate the public with mind-numbing numbers. Analogy, the traditional device of popularizers, will prove useless in coming to grips with them – what is the point, after all, of saying that a stack of 10 trillion one-dollar bills would reach beyond the moon? Nothing less than a transformation of quantitative thinking will be required to keep up with progress, and the logarithm is just the tool for the job. Thinking in terms of logs – power thinking – the stock-in-trade of cosmologists and elementary particle physicists, will have to become a common habit of mind, lest we all drown in a sea of digits.

Both generally and for the purposes of this book, however, the most profound example of power thinking cropped up in the nineteenth century in connection with thermodynamics, the science of heat, and led, surprisingly, to Shannon's quantification of information. By that route, power thinking, of which the simplest example is bit-counting, came to dominate the science of information.

Consider transmitting information digitally, by means of a string composed of zeroes and ones. An important question for a communications engineer is this: how many different messages can you send with such a string? Intuitively it makes sense that you can send more messages with a long string than with a short one, but for the purposes of science this connection should be made quantitative. Consider, for example, a string of three binary digits.

It is easy to enumerate the eight different messages that could be encoded in this way: 000, 001, 010, 011, 100, 101, 110 and 111. But if the string were ten digits long instead, enumeration would become extremely tedious. It turns out that the logarithm represents a simple way to count the number of messages for a string of ten digits, or indeed for any length.

Returning to the example of a three-digit string, notice that 8 is equal to $2 \times 2 \times 2 = 2^3$. This translates into the mathematical statement: The logarithm (to the base 2) of eight is three. By the same token, a ten-digit string can transmit $2^{10} = 1024$ messages. In fact, for any string the logarithm (base 2) of the number of messages is the length of the string. Turn that sentence around: *The length of the string is the logarithm of the number of possible messages*. In this way the length of a string, determined by simple bit-counting, automatically furnishes a measure of the amount of information that can be transmitted by means of that string, provided you think logarithmically.

11 The Message on the Tombstone

The meaning of entropy

If fame is measured by the company a person keeps in death, then Ludwig Boltzmann was a very great man indeed. He is buried in the celebrity section of the Central Cemetery in Vienna, where three million people lie interred far from the central city, halfway out to the airport. His grave lies just across a well-swept path from the shady grove in which faithful fans from all over the world pay homage to Mozart, Beethoven, Brahms, Schubert, Gluck and no fewer than four members of the musical Strauss clan. Boltzmann died by his own hand in a fit of depression while holidaying in Italy with his family, but it is his life, not his sad end, that is celebrated here. His contributions to science are eminently worthy of his place of burial. As monuments to the human spirit, his discoveries rank right alongside the unforgettable melodies of the composers around him. Only, how do you explain that to a young man from Tokyo, strolling through the cemetery with his Australian girlfriend, in search of the Mozart memorial?

Boltzmann's tombstone doesn't do it. It is about five feet tall, made of white marble and adorned with a handsome bust of his stern, bearded face. It bears his name, the dates 1844–1906 and, above those inscriptions, a single bald statement: $S = k \log W$. Because this crisp little mathematical formula lacks the immediate expressiveness of, say, Niels Bohr's ubiquitous planetary atom or the double helix of DNA, this hieroglyphic cannot deliver its message to the uninitiated. But for a physicist, versed in its terms, it stands as a symbol of one man's astounding insight into the workings of nature. In mathematical shorthand it compresses as much meaning, as many connotations, as much information, into its tiny frame as the fragment of Democritus's writings and the opening bar of Beethoven's Fifth Symphony.

The roots of Boltzmann's formula go back to the middle of the nineteenth century, when he was still a boy – a period that saw both the invention of the concept of energy and the triumphant discovery of its conservation; but for all the power of the first law of thermodynamics, as the principle of energy conservation came to be called in the context of the study of heat, it was still incapable of explaining a phenomenon as trivial as the cooling of a cup of tea standing on a table. Why does heat flow out of the cup? Why can't the tea grow warmer, or even begin to boil, by stealing a bit of the tremendous amount of energy stored in the jiggling of the countless molecules in the surrounding air? Why does heat tend to flow downhill, as it were, from hot objects into cool ones, and never the other way around?

In order to answer that question, the German physicist Rudolf Clausius invented a new quantity, which he called 'entropy' as a counterpoint to energy. Entropy, defined mathematically as heat divided by temperature, is easily measured in the laboratory and has a peculiar property that is not shared by its older sister named energy. When a system is left to its own devices, like a cup of tea left on a counter, the entropy, added to the entropy of the surrounding air, always remains constant or increases; it never diminishes. This 'one-way' tendency is reminiscent of the 'one-way' flow of heat, and indeed that's why Clausius invented it. In this way he was able to announce the mighty Second Law of Thermodynamics: without external intervention, entropy remains constant or increases.

It is easy to see how this law quantifies the commonplace obser-vation of heat-flow from high to low temperatures. Imagine two identical cubical blocks of iron. The first one is warm, say 100 degrees on some convenient scale of temperature, while the other one measures a cool 1 degree. When they are brought together to touch along one face, heat begins to flow from the warmer to the colder. Energy, in the form of heat, is strictly conserved in this process. However, if one single unit of heat is transferred, the first block loses an amount of entropy equal to heat/temperature = 1/100. At the same time, the second block gains entropy in the amount of 1/1 = 1. The net gain in entropy of the entire system of two blocks, gain minus loss, is therefore 1 − 1/100 = 0.99 units. If heat were to flow in the opposite direction (which it can't, without the help of a refrigerator) the calculation of the net entropy change would yield −0.99 units, which would mean that entropy would diminish and that the Second Law would be violated. The effect of Clausius's neat invention was to quantify what we all know from experience: when we run down to the cafeteria to buy a cup of hot tea, as well as a cup of cold orange juice for a colleague, and bring them back pressed together in a paper bag, the tea will cool down and turn the juice lukewarm. It will not, alas, emerge hotter, with the juice correspondingly chilled.

The Second Law of Thermodynamics has proved an indispensable tool for physicists, chemists and engineers who deal with the flow of heat. Through it, they have learned how to define and measure entropy in countless different circumstances, and have developed an intuitive appreciation of entropy's behaviour – one to rival a plumber's understanding that water must either stand still or run downhill. If you want to reverse its flow, you must perform work and expend energy, as those of us who are saddled with basement sump pumps know only too well. By the same token it takes energy to reverse the flow of entropy.

In time, the Second Law assumed 'the supreme position among the laws of Nature', as the English astrophysicist Sir Arthur Eddington put it; but for Boltzmann that was not enough. Entropy is heat over temperature – but why? What does that ratio mean? By the time Boltzmann began to think along these lines, heat had been unmasked as nothing but the total energy of motion of countless atoms jiggling chaotically. Likewise, temperature was, by then, known to be a measure of the average energy of individual molecules, gas pressure had been interpreted as the collective push of molecules against the sides of a container, while mass had been pinned down as the sum total of the masses of all the atoms in the system. In other words, all the quantities needed to describe the phenomena of heat were understood in molecular terms *except* entropy. The nature of entropy was as enigmatic in the Victorian era as information is today.

What follows is not a historical account of how Boltzmann actually solved the riddle of entropy, nor how his pioneering work was completed and extended into a powerful theory by the American theoretician John Willard Gibbs. Instead, I will tell a little story to illustrate the kind of examples of one-way processes that led Boltzmann to his famous formula.

Imagine that you have four pairs of identical, closed, glass vessels shaped like litre bottles of milk with flat, rectangular sides. Fill one of the vessels with some coloured gas (say reddish-brown bromine, but don't inhale it, because it is noxious) and evacuate its partner. Now connect the two bottles by means of a hose, and open the passage between them. You know what will happen: the bromine gas will expand rapidly to fill the entire available space. Since the gas molecules don't hit each other very often, and bounce elastically off the walls like tennis balls, they all keep their original speeds, but spread out to occupy twice as much space as before. So the temperature of the gas will remain the same, while the pressure drops, and the volume occupied by the bromine doubles.

Fill the next pair of bottles with water, but make one as hot as you can make it, and the other one very cold. Press the bottles together along their flat sides. Here something analogous happens, albeit much more slowly. The water molecules keep their places, but by colliding with their neighbours they share their speeds – the fast ones slowing down, and the slow ones speeding up – until the entire volume of two litres settles down at a common, intermediate temperature.

Fill the third pair of vessels with two different types of gas, say bromine and air, at a common pressure and temperature. Connect them as before, and open a passage between them. The two gases will begin to intermingle at a speed intermediate between cases one and two, more slowly than a gas expending into a void, but faster than two bottles coming into thermal equilibrium. Eventually both bottles will be filled with a pale mixture of bromine and air.

Finally, fill the fourth pair of bottles with bromine at the same temperature and pressure. When a passage is opened between them you will not see any change at all.

Boltzmann was well aware of these four observations, and many other similar ones in addition. In each case, something irreversible happens, something like glass breaking or water running downhill, something that it would require a special effort to undo. There are telling similarities, as well as differences, between the four experiments. In the first, gas molecules keep their speeds but spread out; in the second, water molecules more or less retain their positions but share their speeds; in the third and fourth, the molecules intermingle. In all cases energy is conserved. In the first three cases, according to Clausius, entropy increases, while in the last example the entropy remains constant. What quantity, defined strictly in terms of atoms and molecules, displays those same tendencies and can therefore be considered as a candidate for a molecular model of entropy?

At this point in the argument Boltzmann was struck by that lightning bolt of inspiration that earned him his place among the immortals. He realized that whatever it was that increased had to do with the *arrangement* of the molecules – not the energy, or volume, or temperature, or chemical composition of the system. Boltzmann noticed a crucial difference between the arrangements of the molecules before and after the two bottles were allowed to interact with each other. If you throw atoms at random into two bottles, they will probably not just fill one, but both equally. If you distribute a total amount of heat energy – by way of putting atoms in motion – between two bottles, it is unlikely that all the fast ones will end up in one bottle and all the slow ones in the other. If you toss two different kinds of molecules into a vessel consisting of two connected bottles, they are unlikely to sort themselves out spontaneously, with one type in one bottle, and the other one in the other bottle. In other words, Boltzmann saw the difference between the initial and final arrangements of the molecules in this light: a random process was more likely to produce the final arrangement than the initial arrangement. Accordingly, he set out to relate entropy to likelihood.

This kind of reasoning was revolutionary. Instead of trying to account for certain, definite properties of particles, such as positions, speeds, weights and sizes, as scientists had been doing since the time of Newton, Boltzmann opened a discussion about random processes, likelihoods, probabilities. Physics would never again be the same.

Our last little experiment, the mixing of two identical gases, by the way, is a sleeper. Despite appearances, the molecules of the two vessels must also wander into each others' spaces, just as they do when the gases are different. Accordingly, the probability of producing the final, mixed arrangement should also increase as if the gases were different, but an outside observer has no way of telling (since all the molecules are identical, it is impossible to determine their degree of intermingling by chemical or physical means). The fact that the entropy does not change in this case gives a sudden and powerful hint that entropy is not so much an absolute property of a body, like weight, or volume, or composition, but that it must be related somehow to what we know about it – to the information we can assemble about it.

Boltzmann's discovery can be recast in a way that avoids the tricky concept of probability. The probability of finding a particular arrangement of the components of a system is related to the number of ways in which the system can be prepared. Suppose you throw two dice and count (i.e. measure) the total number of points that come up, regardless of how they are distributed. There are thirty-six different outcomes, but only one of those is a twelve. The probability of throwing a twelve is therefore 1 in 36. On the other hand, a seven can be thrown in six different ways, so the probability of getting seven is 6 in 36 or 1 in 6. Instead of talking about probability, then, one can equivalently talk about the *number of ways*, which is shorthand for 'the number of ways in which a system can be prepared without affecting the measured properties of the system'. In the example of the dice, the *number of ways* for twelve is one, and for seven is six.

When Boltzmann tried to make the connection between entropy and *number of ways* specific, he ran into a forbidding mathematical problem. Chemists, who use entropy as an everyday tool of their trade, know that entropy is a simple additive quantity like mass and energy: combine two identical volumes of gas, and you get twice the mass, twice the energy and twice the entropy of one vessel. But the *number of ways* is multiplicative: one die can fall in six ways, two dice in $6 \times 6 = 36$ ways. So entropy and the *number of ways* cannot possibly be the same thing!

At this point it may not surprise you to learn that Boltzmann reached for the logarithm, with its marvellous ability to turn multiplicative quantities into additive ones by simply counting digits. But since entropy is measured in units of heat over temperature, while the log is a pure number, he first needed a constant of proportionality – a fudge factor – to complete his law: *entropy is the log of the number of ways, times a constant with the appropriate units*. When Max Planck, the father of quantum mechanics, later wrote this equation in mathematical notation, he chose S for entropy in order to distinguish it from energy, k for the constant (*Konstante* in German), and W for the number of ways. $S = k \log W$ – the inscription on Boltzmann's grave.

A memorial more lasting than marble, the number k came to be called Boltzmann's constant. Numerically it has a very tiny value – 10^{-23} in conventional units – and therein lies a puzzle. The other side of the equation, the entropy, is measured in the laboratory with thermometers, rulers, scales and pressure gauges, so its value comes out to be a normal, everyday sort of number, like 1, or 0.75, or 13.9 – not particularly tiny or huge. In order for Boltzmann's famous formula to work out, the log must therefore be large (around 10^{23}) in order to balance the minuscule size of k; but in the last chapter we saw that one of the principal advantages of the log was that it is usually a nice, intuitively comprehensible number. Why, then, does it turn out to be so huge in this case?

The surprising answer comes from an estimate of W, the number of ways, for a typical system. For example, what is the number of ways of rearranging the molecules in a bottle full of air at room temperature, without changing any of the measured attributes of the bottle, such as its volume, pressure or temperature? W has been estimated to be not just large, not just huge, but positively immense. It is so big that you couldn't write it out in your lifetime. W turns, out to be something like ten to the power 10^{23}, which is to say a one with 10^{23} zeroes trailing after it. Written out in full, W would reach to the stars. It is a number 'so far from ordinary intuition [that it] elicits intense revulsion in some people, and immoderate enthusiasm in others'.

The significance of Boltzmann's interpretation of entropy transcends the fact that it furnished, for the first time, a mechanical model of this useful quantity. Although mechanical models are as old as Democritus, W, the number of ways, has a strange subjective quality that was entirely new to physics at the time it was introduced. It is not just the number of ways a system can be rearranged, but more specifically the number of rearrangements consistent with the known properties of the system. Known by whom? Measured by what observer? Suppose that some superhuman being with an unimaginably potent memory could measure and record all the positions and speeds of the molecules in a bottle of air. In that case W would be exactly one, consistent with the memorized data, and since the log of one is zero (that's where the log curve crosses the horizontal axis on its way from minus infinity to plus infinity) the entropy of the system comes out to be zero. Of course in common practice such a being is unavailable, the molecular states of motion are unknown, and the system can be rearranged in a Gargantuan number of ways. By this reckoning entropy is not an absolute property of a system, but relational. It has a subjective component – it depends on the information you happen to have available.

Boltzmann understood this connection and made it more specific. He pointed out that since the value of entropy rises from zero, when we know all about a system, to its maximum value when we know least, it measures our ignorance about the details of the motions of the molecules of a system. Entropy is not about speeds or positions of particles, the way temperature and pressure and volume are, but about our lack of information.

Let's try this interpretation on our four sets of twin bottles. What happens when bromine escapes from one vessel and fills two? We didn't have much information about the molecules to start with, but we did know that they were restricted to one specific volume. As soon as the gas expands, we lose that bit of information – our ignorance about the position of any specific molecule increases, and hence the entropy increases along with it. In the experiment with warm and cold water, we start out with the knowledge that the faster molecules are in one bottle, and the slower ones in the other. After the two bottles are brought into contact, and the temperature settles down at some intermediate value, we lose that information, paltry as it seems, and so the entropy rises, in agreement with the Second Law. In the third case, in which different molecules get all mixed up, we again end up knowing less about their whereabouts – but in case four, the mixing of identical gases, we remain no more ignorant afterwards than we were beforehand. By deriving the Clausius formula for entropy in cases such as these, and many others besides, Boltzmann was able to show that his interpretation of entropy as a measure of missing information was indeed correct.

Boltzmann's measure of information was simply W, the *number of ways* of rearranging a system. When that number is large, our ignorance is large; when it is small, our ignorance is correspondingly small. In this roundabout way – by identifying entropy with missing information – Boltzmann hurled the concept of information into the realm of physics. That this connection should make use of the notion of 'arrangement' or 'order' is not surprising in view of the list of synonyms of 'form' we encountered in chapter 3, which included both of those words.

12 Randomness

The flip side of information

By now, abstract though the concept is, *information* should be emerging from the pages of this book as a key to understanding principles as diverse as thermodynamics and quantum mechanics in physics, and the operation of heredity in biology. By way of the systematic study of communications it has now become a part of applied science. Today it has its own branch of engineering called information technology, and, further, stands poised to join the concept of *energy* as a unifying thread that runs through all of science, linking diverging branches and unifying the foundations of the whole enterprise.

The two definitions of information we have encountered, proposed respectively by humanists and scientists, suffer from opposite maladies: where the first seems too broad to be of use, the second is surely too narrow. Nevertheless, we will continue to pursue the narrow definition, guided by the spirit of reductionism, both petty and grand, which has hovered over science since its inception in classical antiquity.

Shannon's technical definition of the information content of a message – the number of digits when the message is written in the binary code of the computer – doesn't distinguish between sense and nonsense. A significant sequence of digits like my credit-card number carries the same amount of Shannon information as a random string of the same length. Perhaps, then, the problem of information could be attacked from the other end – for the opposite of meaningful information is random information. If random elements were first eliminated from a message, its meaningful content might be more readily recognized and measured, and we'd be making progress towards understanding information. But what is random?

Randomness is defined as lack of form or pattern, so the adjective 'random' means 'formless, patternless, and therefore unpredictable'. The shape of a Rorschach inkblot (or at least half of one) appears random, as does the order of cards in a well-shuffled pack – and so, in terms of electric raindrops of information, it would seem reasonable to expect a random string to be one whose binary digits follow one another in an irregular and orderless manner. The string 00001001001101000000 seems like a good place to start. It certainly has no obvious pattern to it – but is that enough to make it random? Could it say, represent the tosses of a coin, assuming that 1 means heads and 0 tails? Counting reveals that tails outnumber heads by 14 to 5 – an unlikely proportion. Maybe the penny is biased!

Thus a clarification suggests itself: 'Zeroes and ones must appear (approximately) equally often.' So, is 01010101010101 . . . random? 'No,' you grumble, 'I've been tricked. Zeroes and ones are certainly balanced, but the sequence is much too regular. In fact, it is predictable, whereas a random sequence is not!' OK, so our requirement for randomness was only a first step. The sequence 01010101010101 does indeed pass the most primitive test for randomness, but suffers from a new problem. At first glance this string seems perfectly balanced, but if you look a little closer, its heavily biased nature quickly becomes apparent. Notice that while zeroes and ones occur equally often, and that the pairs 01 and 10 are well represented in it, other possible combinations of digits, such as 00 and 11, never appear at all! So an improved requirement for a random sequence might be: 'The digits 0 and 1 must occur equally often; the four pairs 00, 01, 10, 11 must also occur equally often; the eight triplets 000, 001, 010, 011, 100, 101, 110 and 111 must occur equally often; and so on for all possible subsequences.' According to this new condition, our string of alternating zeroes and ones is clearly not random.

Since every finite sequence of zeroes and ones represents a real number written in binary code, sequences might just as well be called numbers. Mathematicians call a number that passes the new, more stringent test for randomness 'normal', in the sense that it doesn't exhibit a suspicious preference or antipathy for any subsequence. There are obviously many different normal numbers: once you have one, you can always create others by just switching around a few of its digits without violating the requirement for randomness.

Amazingly, although it is known that most numbers are normal in a theoretical sense, the first explicit example of a normal number was not produced until 1933 when David Champernowne, then a twenty-one-year-old undergraduate at Cambridge University, found one. He pointed out that if you start with zero, proceed to one, and then systematically string together all four pairs and then all eight triples, and so on through all possible combinations, you get a sequence (called Champernowne's Number), which must, by construction, contain all possible patterns in the right proportions. A related normal number in the decimal system is $c = 0.123456789101112...$, which contains the right proportions of all integers with single digits, double digits, triple digits, etc. This is known as Champernowne's Constant, and has the advantage over Champernowne's Number of being both small and finite, so it can be plotted on graph paper – just a little bit to the right of the origin.

David Champernowne, nicknamed 'Champ', was a college classmate and lifelong friend of Alan Turing – one of that genius's few intellectual peers. They played chess together and after the Second World War even created a rudimentary computer chess program called Turochamp, which managed to beat Champ's wife Wilhelmina, then a raw beginner. Champ went on to a distinguished career in economics and statistics, and died in September 2000 at age eighty-eight.

Champ's normal number in its binary form is a fabulous object. Using Morse code, or some other translation of zeroes and ones into typographical symbols, it can be transformed into a string of letters, spaces and punctuation marks. Since every conceivable finite sequence of words is buried somewhere in the string's tedious gobbledygook, every poem, every traffic ticket, every love letter, and every novel ever written, or ever to be composed in the future, is there in that string. All of Shakespeare appears in multiple versions, some with and others without misprints. *War and Peace* is in there, of course, but so is a French translation of the US Constitution in hexameter form. You may have to travel out along the string for billions of light years before you find them, but they are all in there somewhere. Thus Champ's miraculous number, a mere dot in its decimal version, surpasses in compactness Jorge Louis Borges's vast *Library of Babel*, which also contains all possible combinations of letters, but stores them on an infinite row of bookshelves. As a bibliophile, I am thrilled and vaguely disturbed by the notion that I can replace all my books, plus those in the Library of Congress and the Bibliothèque Nationale in Paris, with a dot painted on my thumbnail, just a little bit to the right of the edge of its cuticle.

But does 'normal' mean random? Is Champ's Constant random? If you happened upon a segment of it somewhere far from the beginning, would you still be able to recognize its simple, rigid pattern, or would you be justified in calling it random?

The definition of randomness required to answer this question has not been perfected yet, but an important step forward was inspired by the computer revolution. In the 1960s several mathematicians – principally the Americans Gregory Chaitin and Ray Solomonoff and the Russian Andrei Kolmogorov – linked the notion of randomness to another property of numbers they called 'algorithmic complexity'. They noticed that the string 010101010 . . ., which, as we established earlier, is not even normal, and thus not a candidate for randomness, also looks pretty simple. To refine and quantify this intuitive remark, and to remove the subjective element from the terms 'simple' and 'complex', they imagined a computer programmed to produce the number in question. Roughly speaking, they then defined the complexity of a number as the length of the most succinct program, or algorithm, that can produce it.

Three remarks are needed in order to flesh out this cryptic definition. The inventors of algorithmic complexity added the stipulation of succinctness in order to strip away extraneous or redundant steps – the electronic equivalent of humming and hawing – that can always be added to any program to make it longer. Notice that the requirement of brevity conforms neatly to the injunction of Ockham's razor: briefer is better. If algorithmic complexity is ever to play a role in physics, this provision builds in Ockham's razor – a basic principle of science – in a fundamental way. The second remark concerns the worry about different machines performing operations in different ways, and thereby possibly leading to different values of complexity for the same recipe. Fortunately, Alan Turing had invented a hypothetical standard machine that could reproduce the work of any conceivable computer. The use of Turing's universal engine in the definition of complexity solves the problem of dependence on a particular machine. Thirdly, while programs may be written in English, computers ultimately translate them into a machine language consisting of zeroes and ones. Those are what's counted to arrive at the 'length' of a program.

Bearing these qualifications in mind, the program for generating 0101010101 ... might read: 'Type 0 and 1 alternately a thousand times' – a short command, so the complexity of that number is small. By the same token, consider 14159265 ..., which I used as an example of a number that looks random, but really carries a lot of information when it is recognized as the fractional part of pi. This sequence, which has now been calculated to billions of digits by supercomputers, can be generated by a computer program that evaluates a formula with a simple pattern such as pi $= 4 \times (1 - \frac{1}{3} + \frac{1}{5} - \frac{1}{7} + \frac{1}{9} - \ldots)$. Consequently the complexity of pi is also very small.

Yet for all its appeal, the definition of algorithmic complexity suffers from a near-fatal defect. Say you are confronted with a long string of zeroes and ones. What is its complexity? If you can spot a pattern, or you happen to know how it was generated, the answer is straightforward enough, but what if you don't? How do you know that there is not a computer program somewhere out there in the universe that generates this number, and none other, in a series of simple steps? The answer is clear: you don't! The smell of subjectivity clings to the mechanical definition of complexity as stubbornly as it sticks to the definition of information.

That the idea has any merit at all derives from the fact that in many circumstances a lot is known about the origin of the string from the context of the discussion. If, for example, the data being scrutinized for their information content refer to the positions of gas atoms, then it would seem reasonable to include the specification of their container in the program that generates them, but nothing else. If a string of numbers describes the flow of water through a pipe, the laws of fluid dynamics provide constraints, but no other correlations need to be considered. In general, any additional knowledge that is available about the data in question narrows the search for hidden patterns, and provides a context for the definition of complexity; but it should be safe to assume that gas atoms or drops of liquid in a pipe don't arrange themselves according to some mysterious algorithm.

With these preliminaries out of the way, we can finally return to the definition of randomness. *A binary number is random if its complexity is equal to the number of its digits*. This mouthful means that there is no short recipe for producing the number – the only way to generate it is to write it out in its totality – it harbours no regularities that would simplify its specification. By this definition 010101010 ... is neither normal nor random. Champ's Number is normal but not random, because the instructions for generating it are quite simple. Even pi, which looks normal and patternless out to billions of decimal places, is far from random: its complexity is measured by the number of digits of the short program needed to define it. On the other hand 84029587614390, which I just tapped out on my keyboard, cannot, as far as I know, be described by a shorter program, so it's random.

Actually, come to think of it, I'm not so sure it really *is* random. Claude Shannon invented a 'mind-reading (?) machine' [his question mark, not mine] for playing penny-matching. It was basically a primitive computer programmed to guess a human opponent's next choice of heads or tails by recognizing trends in the previous choices. If there were no trends or hidden patterns, the computer should only succeed 50 per cent of the time. In fact, however, after a colleague actually built this machine, it won between 55 and 60 per cent of the time, showing that people are not very good at generating true randomness. The irrepressible Shannon didn't rest there, however. He built an improved version of the machine himself, which beat his colleague's.

A more consequential example of the human inability to produce random sequences actually helped British cryptanalysts to crack the notorious Enigma code. In the German army's version of the code, the operator was supposed to fix the machine setting for the day by transmitting a set of randomly chosen letters; but under the pressure of war, many soldiers succumbed to the temptation to use their own initials, or those of a girlfriend, or actual words, instead. Their human frailty betrayed them.

The irony of randomness is that any recipe for generating a random number on a digital computer is bound to fail. ('Anyone who considers arithmetical methods of producing random digits is, of course, in a state of sin,' preached the computer pioneer John von Neumann.) The random-number generators included in commercial statistical software packages rely on computer programs that are shorter than the strings they can generate. For some of them, hidden structure can be revealed by plotting their outputs as points in a three-dimensional box. At first the resulting dots seem to spread out uniformly, as expected; but in a paper entitled 'Random Numbers Fall Mainly in the Plane' the mathematician George Marsaglia showed that when looked at from some cleverly selected, special directions, the points actually form a series of discrete parallel planes. (They remind me of parts of the Black Forest of Germany, which look perfectly natural from a distance, but reveal their cultivated origins to the passing driver who happens to catch an unexpected glimpse of their straight and perfectly parallel rows of trees from certain special vantage points on the road.) When this egregious flaw of so-called random-number generators was discovered, it caused consternation among the mathematicians, physicists and social scientists who thought that they had modelled chance behaviour with their programs. The software was renamed 'pseudo-random-number generator', and adorned with warning labels.

Producing random numbers is an important cottage industry serving two kinds of applications. Computer simulations of systems that include supposedly random events – from stock market fluctuations and weather patterns to the behaviour of X-rays aimed at tumours – depend on the integrity of input data that are assumed to be unbiased. The second application is a clever mathematical technique called Monte Carlo integration, which is used routinely to compute areas and volumes of complicated shapes in many dimensions. By way of illustration, consider a heart drawn on a 10 cm × 10 cm sheet of paper. What is the area inside the heart? An analytical mathematician might try to find a formula that reproduces the heart-shaped curve, and then use the well-known techniques of integral calculus to arrive at the answer. For the messy curves found in real life this approach is usually impossible. A numerical analyst would, however, reach for the Monte Carlo method. Using a random-number generator to specify their coordinates, she would sprinkle 1000 imaginary dots evenly over the whole sheet. Then she would count the number of these dots that happen to fall inside the heart, and divide it by a thousand. The result is the desired answer, expressed as a fraction of the total area of a hundred square centimetres. It is not an exact method, but since its accuracy improves with each additional dot, the correct answer can be approached as closely as desired.

That example illustrates the need for truly random numbers. If the dots were distributed more densely in the middle than near the edges, or vice versa, the calculation would obviously be flawed. With other kinds of patterns among the dots, the accuracy of the calculation would be compromised in other ways. As simulations and Monte Carlo integration programs have multiplied in science, engineering and economics, the demand for random numbers has grown correspondingly.

If random numbers can be generated neither by humans nor by

computers, where can they be found? Even in this digital age, computer scientists have reached for the equivalents of a monkey with a typewriter, such as noisy electrical circuits, dripping faucets and lava lamps. More practically, though, since nature is believed to behave randomly on the quantum-mechanical level, the latest device for generating random numbers is not a computer, but a laser. Inside a book-sized box a feeble beam of photons is split into two beams, which are directed into two detectors. When a photon triggers the first one, a zero is recorded, while the second one yields a one. In this fashion a reliable random sequence is created. Contrary to Einstein's intuition, not only does God play dice, but He seems to be the only one we can rely on to play fair.

The success of a mathematical definition is measured by its usefulness in solving old problems and its ability to suggest new ones. By these criteria, the notion of algorithmic complexity lay fallow for a long time after its invention in the nineteen-sixties. It played no role, for example, in the frantic search for better pseudo-random-number generators that accompanied the spread of Monte Carlo calculations; but in recent years complexity has finally begun showing signs of maturing into an effective mathematical tool. Once the ice was broken by its application to a problem in pure mathematics, a growing number of experts in computer science, applied mathematics and even economics have started to consider it for help with the issue of randomness.

An obscure mathematical theorem proved in 1999 illustrates the uses of algorithmic complexity. The Dutch mathematician Paul Vitanyi, together with Tao Jiang of McMaster University and Ming Li of the University of Waterloo, both in Ontario, decided to tackle a particular version of a venerable geometrical problem. Consider a square, ten centimetres to a side, drawn on a blank piece of paper. Sprinkle a hundred dots randomly over this square, and connect all the dots by straight lines. The square is now covered with a fine web of triangles, like the wrinkles on an old lady's cheek. By measurement, determine the area of each of those little triangles, and focus on the smallest one you find. It is obvious that if you increase the number of dots, the area of that smallest triangle will, when averaged over many random sprinklings, become even smaller; but mathematicians are curious folk, and want to know by *how much* the average shrinks. If, for example, you double the number of dots, will the average area diminish by a factor of two – or four – or eight?

This is the sort of arcane conundrum that delights mathematicians, so over the course of the past fifty years considerable ingenuity has been applied to various versions of it. Paul Erdós, the peripatetic genius we met in chapter 9, contributed his share to the effort in the nineteen-eighties; and now Vitanyi and his colleagues, by defining randomness in terms of complexity, were able to solve it. It turns out that the average size of the smallest triangle varies inversely as the cube of the number of dots, so doubling the dots makes the smallest triangles eight times smaller. Although the triangle problem is surely not terribly important to anyone but a mathematician, its solution succeeded in the express purpose of its authors to introduce the usefulness of algorithmic complexity and randomness 'to a wider audience by exhibiting a new example application'.

The techniques of complexity theory may find applications in both information theory and physics. In order to measure the information content of a message, for example, it might turn out to be useful to compress the message before counting its digits. This can be achieved by finding the shortest program that can reproduce the message. Thus a second copy of a book would not double the information content of the original, but only increase it by the length of the trivial program: 'Print twice.' In this way Shannon's overly broad definition of information would be reigned in.

A potentially important application of algorithmic complexity to physics was proposed by Wojtek Żurek of the Los Alamos National Laboratory in New Mexico. In order to rid Boltzmann's definition of entropy of its troublesome element of subjectivity, Żurek suggested an almost imperceptible modification of it. Recall that entropy is a measure of missing information about a system. It therefore depends on what an observer happens to know: a smarter being has more information, is missing less, and thus assigns a lower entropy to a system than a more limited creature. To render entropy more objective, Żurek recommended adding a measure of *recorded* information to that of missing information. The sum of the two remains constant – if you remove data from one column, it reappears in the other. The observant creature thus becomes redundant; only the entries in its notebook or computer memory matter.

But how to assess the amount of recorded information? Żurek chose algorithmic complexity as the most natural measure. Accordingly, his new, improved entropy consists of two portions: the conventional entropy as measured by the formula on Boltzmann's tomb, plus a piece that is normally inconceivably tiny, and accounts for the algorithmic complexity of the listing of recorded knowledge about the system. A mathematical description of the size and shape of a vessel containing a gas might be a typical item in the list, while missing information includes the coordinates of a vast number of atoms. Notice that in the hypothetical case that every position and every velocity of every atom is known, the Boltzmann entropy of the system is zero, but the added term – the length of the description of what's known, in binary code – will be huge, bringing the total entropy back to its previous value. After a hundred years the reek of subjectivity has finally been lifted from the Second Law of Thermodynamics.

In spite of its cogency, Żurek's improved entropy has not gained much support. Under normal circumstances Boltzmann's old formula works sufficiently well that any modification can be ignored; but as the atomic systems that we can manipulate and analyse become smaller, and as our information about them grows by leaps and bounds and can be stored in ever more powerful computer memories, it may eventually happen that Żurek's addendum will become large enough to be measurable. Then people will have to sit up and take notice.

The valiant attempt to define information from the vantage point of its equally elusive antithesis, randomness, has not succeeded yet. Some techniques that arose in the course of this endeavour, such as algorithmic data-compression, may well play a role in a future theory of information, but to date the reigning candidate for measuring information remains Shannon's bit counting. The time has come to examine it more closely.

13 Electric Information

From Morse to Shannon

The date and time of the birth of electric information are certain: Friday, 24 May 1844 at nine forty-five in the morning. The place, however, is less so, for the happy event occurred not at a point, but on a line connecting two points – a pair of wires from the Supreme Court Chamber in Washington, DC, to an office in Baltimore, forty-one miles away. In the middle of the courtroom, amid a tangle of wires, Samuel Finley Breese Morse sat hunched over a mysterious brass box, surrounded by a curious crowd of congressmen elbowing each other to get a better look over his shoulder. He was understandably anxious, because a year earlier Congress had awarded him the substantial sum of $30,000 to mount a historic experiment, and if it failed, his bold project would be doomed. Urgently he fiddled with the complicated screws and levers needed to generate a series of electrical dots and dashes. In the style of the time the message he composed, which had been selected earlier by the young daughter of a friend, was the Biblical exclamation: 'What hath God wrought!' A few minutes elapsed while a colleague in Baltimore decoded the signal, and then, to the astonishment of the crowd and the relief of the inventor, the same message, faithfully copied and re-encoded, arrived back in Washington.

Three days later the *New York Daily Tribune*, presumably informed by mounted messenger, triumphantly described the event, and declared that the miracle of the annihilation of space had been accomplished. The support of the US Congress, and the publicity resulting from the spectacular demonstration, touched off the furious growth of telegraphy. Private enterprise rushed in to capitalize on what the public purse had subsidized. Of course, Morse had not been the only inventor to try his hand at applying new discoveries in electricity and magnetism to the problem of communication. A number of rivals in France, Germany and England had been hard on his heels, and in some respects even ahead of him, but eventually his system, which he defended vigorously in countless patent suits and priority battles on both sides of the Atlantic, had triumphed over theirs. 'It would not be long', Morse prophesied, 'ere the whole surface of this country would be channelled for those *nerves* which are to diffuse, with the speed of thought, a knowledge of all that is occurring throughout the land, making, in fact, *one neighbourhood* of the whole country.' (A hundred years later Marshall McLuhan, the prophet of the information age, would echo the obvious metaphor: 'Today, after more than a century of electric technology, we have extended our central nervous system itself in a global embrace, abolishing both space and time as far as our planet is concerned.') In fact, Morse underestimated the range of his invention. Within twenty years telegraph cables crossed not only the continent, but the Atlantic Ocean as well to create the very first world-wide web. Information was its staple, the Morse code its language.

At the time of the demonstration in Washington, Morse, a Massachusetts Yankee born near Boston and educated at Yale, was fifty-three years old. The telegraph, though years in gestation, was by no means his first accomplishment. In fact, Morse's career had started out in a very different direction. Until just seven years earlier he had been a respected portrait painter, with over three hundred works to his name, of which several now hang in the principal museums of America; but in 1837, at the peak of his powers, he had suddenly stopped painting. For the rest of his life he devoted himself to the brilliant idea that had first come to him during the enforced leisure of an ocean voyage from Europe back to America. The reasons for his abrupt change of heart are revealing.

Its proximate cause was bitter disappointment at being passed over for a commission to paint a mural for the rotunda of the Capitol in Washington; but this setback, by itself, would not have been enough to deter a single-minded artist. Morse was, in fact, driven by other urges besides devotion to art. For one thing, he hungered for fame and fortune. For another, he was consumed by an abiding zeal to preserve the nation's culture from the levelling force of the prevailing populism. At heart he was a teacher and a reformer, and he was prepared to use any available means in the pursuit of his ideological goals.

The two paintings to which he had devoted the greatest effort were *The House of Representatives* (1822) and *The Gallery of the Louvre* (1831–33), two vast, wall-sized canvases. The subject of the first is the machinery of government and those who operated it; over a hundred personalities, painted from life, are shown assembled for a lamp-lit evening session of the House. In the second, Morse himself takes centre-stage, tutoring a young girl as she sketches, against a backdrop of the greatest European art then exhibited in the Louvre, including da Vinci's *Mona Lisa* and Rembrandt's *Head of an Old Man*. The motivation for both works is obvious: they were not meant to hang in homes or museums, nor to delight the elite; rather, their purpose was to instruct the public. Supported by detailed explanatory notes for the viewer, they carry educational messages – the first about democracy in action, the second about high culture.

Unfortunately for Morse, both paintings were roundly ignored by audiences in New York and Boston. Whatever their artistic merit might be, in their own time they were failures. Chagrined and bitter at being rejected first by the public and then by Congress, Morse renounced art and, after a short and unsuccessful campaign for public office (on an anti-abolitionist, anti-Catholic and rabidly xenophobic platform) turned to telegraphy. In a way, he thereby remained true to his original purpose. For, in the hurly-burly of mid-nineteenth-century America, its development suggested a new and thoroughly modern way of conveying information, of communicating and instructing – something that painting and sculpture had been doing since prehistoric times.

Before one could think of reaching the public, however, or of harnessing the lightning speed of electricity and the power of magnetism, one had to start with a system, and the simplest possible proved to be a wire and a linear sequence of dots and dashes – zeroes and ones. Thus, one of the first problems Morse tackled on his way to designing a telegraphic system was the question of coding: how do you translate verbal information into electrical signals, which in turn make visible or audible marks for the recipient? His initial solution to this fundamental problem of information technology turned out, in retrospect, to be far more efficient than his ultimate code, but too clumsy in actual practice. Since the telegraph grew out of the science of electricity and magnetism, it was natural to turn first to numbers, the alphabet of science. Numbers are easy to record as clicks of a counter or scratches of a pen. These, in turn, could be generated by electrical currents activating electromagnets. (One of the principal differences between Morse's design and those of his European competitors lay in its effective use of electromagnets, which had been perfected by the American physicist Joseph Henry.) Accordingly, Morse's first code was numerical.

In 1837 he assembled the numbers he assigned to specific words and phrases into a special dictionary. Telegrams would consist of nothing but lists of numbers, which the recipient would decode by reference to the dictionary – a system that is clearly capable of an impressive level of data compression. If, for example, the sentence: 'The eastbound train will arrive three hours late,' is assigned the numeral 3, then only three clicks (or two bits, since 3 is represented by the binary code 11) are required to transmit that eight-word, fifty-symbol message. It would be a hundred years before the theoretical efficiency of Morse's original system would be recognized. In his time, however, the human effort involved in looking up numbers in a big book and transcribing them manually proved to be an insurmountable bottleneck.

Undeterred, Morse pushed on to invent the alphabetic code that bears his name. While infinitely more flexible than the numerical code, its messages are much longer. This trade-off led inevitably to the international style called 'telegraphese', in which a message might read: 'Eastbound delay 3 h stop.' But even this compact sentence, transmitted letter by letter, is relatively long and costly, and needed to be kept as brief as possible. Time, after all, is money. So Morse confronted the problem that is at the root of modern information theory: what is the most efficient way to choose symbols, composed of dots, dashes and spaces, for the letters of the alphabet? The ingenious way in which he answered this question, and the way it was tackled a century later, illustrate in microcosm the difference between engineering and science, between practice and theory. Then, it was a triumph of Yankee resourcefulness; today it bears witness to the power of mathematics.

The principle is obvious: the most efficient code assigns short symbols to the common letters, and long symbols to the rare ones. But what is common, and what is rare? What is the order of the frequency with which letters appear in English? Cryptographers need this information, as Edgar Allan Poe demonstrated in his popular story 'The Gold Bug' a year before Morse's experiment in Washington. One way to gather such statistics is to select a text, and simply count the number of times each letter appears, but while this method works well for the three or four most frequent letters, it becomes successively less reliable for the uncommon ones, such as Q, X and Z, unless the reference text is extremely long. Morse's pragmatic solution, which he hit upon five years before the appearance of Poe's story, was quicker: he walked into a newspaper office and counted the number of letters in each compartment of the printer's type box. Presumably decades of experience had reduced its contents to an efficient compromise between supply and demand. Since he found more Es than any other letter, E is represented by a single dot, followed by T which merits a dash. X, Y and Z, on the other hand, whose compartments in the type box were relatively empty, drew four symbols each.

The theory that would vindicate Morse's rough-and-ready method a century later was devised by the American mathematician Claude Elwood Shannon. Shannon, who was born in Michigan, earned his PhD at MIT and worked at the AT&T Bell

Telephone Laboratories in New Jersey for fifteen years before returning to MIT to teach. He died in February 2001 at the age of eighty-four, laden with honours and revered as the legendary founding father of the cyber age.

In personality, Shannon was the polar opposite of Samuel Morse. He was self-effacing, carefree, retiring and cerebral where Morse was self-important, dour, combative and practical. Both men overflowed with energy and determination, but because Shannon tasted success early in life he never acquired the bitterness Morse was able to shed only at an advanced age when he had finally captured the fame that had eluded him for so long. Their approaches to science differed in the way painting (Morse's art) differs from music (Shannon's): while painting tends to be holistic and synthetic, and is said to be right-brained, music is more analytical, focused on details, and supposedly left-brained. A painting presents itself to the eye in one go, whereas a composition is heard in a linear fashion, one note at a time. Thus Morse and Shannon embody complementary approaches to a technical problem.

The quality that distinguished Shannon was playfulness. Like many scientists he was fascinated by games, puzzles and tricks, but unlike most people, once he got a hold of one, he would persist until he had mastered it and discovered its mathematical essence. To borrow a phrase from James Clerk Maxwell, who shared the same passion, Shannon never stopped until he understood 'the "go" of it'. A reporter who visited him at home after retirement found him surrounded by marvels like the mind-reading (?) machine mentioned in the previous chapter, innumerable normal and strange musical instruments, a collection of chess-playing computers, a petrol-powered pogo stick, a two-seated unicycle, a hundred-bladed penknife, and a computer called THROBAC that calculated in Roman numerals (which was similar in principle to the electrical Marchant desk calculator that introduced me to the art of computing half a century ago). Shannon's most enduring hobby was juggling, which he practised as a young man at Bell Labs while gleefully riding his single-seater unicycle through the quiet, night-time halls. He built an illusion called the 'No-Drop

Juggling Diorama', and wrote a learned treatise on the 'Scientific Aspects of Juggling', complete with poetic epigraphs, historical references reaching back to 2040 BC, schematic diagrams, mathematical theorems, and the design of a diagnostic instrument called a 'jugglometer'. Unlike Morse, who was driven by the search for recognition, Shannon maintained: 'I've always pursued my interests without much regard to financial value or value to the world. I've spent lots of time on totally useless things.'

In science, though, as biologist François Jacob remarked, seemingly insignificant puzzles can lead to deep insights. This certainly proved to be the case with Shannon's investigation of the efficiency of communications channels. Its success was largely due to the care with which he defined and delimited the problem. Figure 1 of his monograph sets the stage: five boxes in a row are connected by arrows from left to right, and labelled, in succession, 'Information Source', 'Transmitter', 'Channel', 'Receiver' and 'Destination'. (There is also a sixth box off to the side, ominously marked 'Noise' and connected to the Channel, but I'll come back to that later.) This simple picture inspired my own questions in chapter 1: 'What mediates between the atom and the brain? What agency originates in the atom, or, for that matter, anywhere in the material world, and ends up shaping our understanding of it?' My answer, 'information', also figures in Shannon's diagram.

The opening of the second paragraph of his seminal 1948 paper 'A Mathematical Theory of Communication' – a work variously likened to the Magna Carta, Newton's laws of motion and the explosion of a bomb – is crucial, and worth recalling:

> The fundamental problem of communication is that of reproducing at one point either exactly or approximately a message selected at another point. Frequently the messages have *meaning*; that is they refer to or are correlated according to some system with certain physical or conceptual entities. These semantic aspects of communication are irrelevant to the engineering problem.

By blithely ignoring the *meaning* of information Shannon succeeded in constructing a complete mathematical theory.

Circumscribed though it is, the theory nevertheless describes a large gamut of phenomena. The information can be written, typed, spoken, sung, played on an instrument, painted, photographed or televised; the channel can be a pair of wires, a light beam, a band of radio frequencies, or any other device for relaying messages; the source and the destination can be people or machines. In order to measure in a consistent way the amount of information flowing through the channel, Shannon, like Morse, had to begin with the problem of coding, and in picking zeroes and ones, those atoms of information, he ended up with a far simpler system. For each choice between the two he chose the word 'bit', and, in order to generalize the concept, he immediately adjusted the original definition so he could use the bit as the unit of information under all possible circumstances.

The key to Shannon's adapted definition of information is a property of the logarithm (to the base 2) that we discovered at the end of chapter 10: *The log of the number of messages that can be sent is equal to the length of the string.* This simple rule prompted Shannon to define the information content of any set of messages as *the log of the number of possible messages*, and to call the unit of information the 'bit'. This definition agrees with the original bit-counting definition, because that's what it was designed to reproduce, but it works even in more complicated circumstances. Imagine, for example, substituting dice for coins. You can communicate with a friend by using the numbers on the faces of a die. There are exactly six different messages that can be conveyed with each die. How much information does each throw convey? According to Shannon's definition, the answer is log 6, which comes out at about 2.585 bits. (The result makes sense, because 6 is halfway between 4 and 8, and its log correspondingly falls between 2 and 3. The specific number can be verified by punching a calculator: $2^{2.585} = 6.000$.) One throw of a die, therefore, is more informative than the toss of two coins, but less than a toss of three – an intuitively reasonable result. In this fashion the definition of the bit is broadened to accommodate fractional values.

The resemblance of Shannon's definition of information (the log of the number of possible messages) to Boltzmann's formula for entropy (the log of the number of ways of rearranging an atomic system) is not accidental. As Boltzmann himself suggested long before Shannon was born, entropy measures the *missing* information about the system – the information one could possibly have, but doesn't. Much ink has been spilled over the significance, or lack of significance, of the connection between information and entropy, but in the end Boltzmann's intuition was reliable: information and entropy are different ways of expressing the same idea.

Shannon's simple information measure sports one last wrinkle. Recall that a bit is the amount of information in the throw of an honest coin, or in the choice between two equally probable outcomes. For the case that the coin is weighted, or that the two choices of a question in the game of Twenty Questions are not equally probable, Shannon devised a formula for information content that involves not only the logarithm, but probabilities as well. Compare, for example, two players trying to guess a town in the US. The first one successively divides the country into two equal portions. With each answer, whether it is yes or no, she narrows down the possible region by half. As we saw, with twenty questions she can divide the country into more than a million patches – with absolute certainty. The second player decides to follow the logic of geography instead. So he begins by determining the state in which the town is located. The answers to his questions are not equally probable, but will be no in forty-nine cases out of fifty. Assuming for simplicity that each state has fifty counties, and that the towns are uniformly spread over the country, the player must use up $49 \times 49 = 2401$ questions just to be able to identify the county with certainty. Shannon's improved formula takes the unequal probabilities of the answers into account, and predicts that the second player requires many more questions than the first in order to elicit the same information. Accordingly, it assigns a smaller information content to each of the second player's questions.

Today, the final, modified recipe for measuring the amount of information, called 'Shannon information', serves as the cornerstone of the vigorous science of information theory. As an industrial tool, this theory's chief aim is to help design machinery for transmitting large volumes of information, cheaply, accurately and quickly over the various channels that have been invented since Morse's primitive wires. It shows, for example, that 'block-coding' (assigning a short word or number to longer phrases) is far more efficient than 'letter-coding' (assigning a different symbol to each letter). Ordinary English requires on average about 28 bits per word. But suppose you insist on restricting yourself to much shorter strings of no more than 14 bits instead. If each string were assigned to a specific word, the system could handle 16,384 (which equals 2^{14}) different words or phrases – plenty for most circumstances. It appears, in short, that Morse's first idea, to transmit messages by means of a numerical code linked to a dictionary of words, was perfectly sound.

More practically, especially since the Morse code is still in use today, one might ask whether its dots and dashes are efficiently assigned. In order to answer this question, it is necessary to take into account the frequencies of letters in English, i.e. the probabilities of 'E' and 'T' occurring, versus 'X' and 'Z'. Shannon's formula does that, in the same way that it accounts for the different probabilities of throwing heads and tails with a biased penny. Before its invention, the problem could not have been solved with mathematical rigour, because a measure of information was missing. Using Shannon's definition, together with the theorems he proved, it has been shown that reshuffling of dots and dashes could improve the Morse code by no more than 15 per cent. Morse's rough-and-ready scheme turned out better than it had a right to be.

Shannon's theory has put Morse's intuitive inventions on a scientific basis. Its real power, however, comes out of the sixth box in his scheme – the one marked 'Noise'. Noise introduces errors, uncertainty, losses and inefficiency into any system. In dealing with robust problems of civil engineering, or the design of big machines, noise is often neglected for the sake of simplifying the problem. In information theory, on the other hand, the noise is often at least as strong as the signal itself, or even stronger, and cannot be ignored. In fact, an information theory that leaves out the issue of noise turns out to have no content.

14 Noise

Nuisance and necessity

The world is a noisy place. Every quiet room is filled with imperceptible sounds, every musical note, no matter how pure, is accompanied by a cacophony of unintended hums and hisses, even the stars glittering in the pristine blackness of deep space would blast our ears with deafening roars if we could get close enough to hear them. Noise is tantamount to unpredictability and uncertainty, so scientists have endowed the word with an extended meaning. Radio static is audible evidence of the jitter in the electrical signal that gave rise to it. Not only radio signals, but all electrical and electromagnetic waves, including X-rays, microwaves and visible light, are afflicted with this tremor. Mixing up the senses, scientists therefore do not hesitate to speak of 'noisy light'. When electronic cables are said to be noisy, the implication is not that they can be heard, but that the currents they carry are disturbed by erratic fluctuations. For the same reason the annoying dusting of 'snow' in every TV image is referred to as noise – even with the sound off.

More metaphorically, when a physicist cuts off a piece of cable for an experiment and finds its length to be too short by a small amount that is still within the tolerances she has established, she is liable to shrug the error off with the remark: 'It's in the noise.' In a hundred tosses of a penny, throwing fifty-two heads instead of the theoretical fifty is said to be 'in the noise'. In fact, all the uncertainties introduced by the imperfections of instruments and observers into every physical measurement, whether it is acoustic, optical, electrical, thermal or mechanical, are called noise.

In terms of information-transfer, noise is a ubiquitous nuisance. It interferes with telephone conversations, corrupts telegraphic transmissions, and limits the accuracy of digital computation. However efficient our technologies become, it is always present, sometimes subtly, sometimes overwhelmingly, but always impinging on our ability to communicate. Its fundamental importance in this respect was powerfully underlined by Claude Shannon – remember that ominously marked little box in figure 1 of his famous essay. Indeed, noise plays such a crucial role in what follows that the entire paper is sometimes referred to as 'Shannon's theory of the noisy channel'.

To fully grasp the impact that noise has on communication it is worth pausing here to consider how things would work without it. Suppose you could send a signal over a wire, or through the air, with perfect accuracy. If you could send a single electrical pulse of arbitrary strength – measured in some convenient units such as millivolts – without incurring any uncertainties along the way from source to destination, all your problems would be solved. You could take a message of any length (the *Encyclopaedia Britannica* would be the gentlest of warm-up exercises for this experiment), translate it into binary code as something like 0010111000 . . ., and send a pulse with a 'height' of exactly 0.0010111000 . . . millivolts to the receiver. There it would be decoded flawlessly, each additional decimal point providing another bit of information, and finally reassembled into a perfect message of arbitrary length.

The same technique could be used with other means of transmission. You could, for example, fashion a ring with an inner radius of precisely 0.0010111000 . . . inches to convey the same message to a friend equipped with a noise-free measuring device and the knowledge of the digital code you employed. So, in a noiseless universe, an infinite quantity of information could be encoded on as simple a shape as an unadorned circle without the need for the clumsy complexity of a book as vehicle.

The fictitious perfect electrical pulse and the golden ring of exquisite design are as intriguing as Champernowne's Gargantuan

Number that incorporates all possible messages. Notice, though, that both the pulse and the ring are considerably neater, because unlike Champ's fanciful construction, they don't hide their messages under infinite piles of garbage.

But it can't be done: nothing in this world is perfect and all messages are necessarily and unavoidably corrupted by noise. No signal can in reality be produced, transmitted or detected with an accuracy of more than a handful of decimal points. Therefore, in order to conquer noise, we must first quantify its effects, which Shannon accomplished approximately as follows. Consider again an electrical pulse and use its strength, or height when plotted along the vertical axis of a time-line, to encode the information. (Morse used its *duration*, rather than its strength, to convey information, but that's another story. Here we look only at its height, not its length.) Say the precision of your apparatus happens to be about one millivolt. This means that the real strength of a pulse measured as 3 could be anywhere between 2.5 and 3.5 millivolts. To quote a value of, say, 3.2306 would be pointless, because the last four digits are in doubt.

The number of different messages that can be transmitted by means of a single pulse that is limited to a maximum height of, say, 6 millivolts, is easily ascertained. Aside from the baseline where no signal is sent at all, one could assign different meanings to values of 1, 2, 3, 4, 5 or 6 millivolts. This particular pulse can thus convey six different messages, as many as a fair die; any attempt to stuff more information into it would fall foul of the limitation imposed by noise. More generally, the number of possible messages carried by an electrical pulse is simply the maximum strength of the signal, divided by the magnitude of the intrinsic noise in the system: six millivolts divided by one millivolt, in this case. The resulting quantity is a pure number which is called the 'signal-to-noise ratio', abbreviated S/N. In accordance with his theory, Shannon proposed to measure the quantity of information in a single pulse not by the actual number of possible messages, which is given by S/N, but by its logarithm (to the base 2), or log S/N expressed in bits. The signal-to-noise ratio can vary from a value of zero (no signal) to a huge number (strong signal, weak noise). If the noise were reduced all the way to zero, S/N would take on an infinite value, and so would its log. In that case a single pulse could carry an infinite amount of information, as we have seen.

If one pulse can carry log S/N bits, then a radio or telephone signal capable of transmitting many pulses per second – say a million – can handle a million times log S/N bits per second. In this way, roughly speaking, Shannon derived his famous formula for the information capacity of a noisy channel that has served as the principal tool for half a century's progress in tele-communications. By measuring the speed with which a channel can convey information, in bits per second, it introduced for the first time a way to compare different techniques for handling information.

So here's that log again! It reared its head in physiology, thermo-dynamics, particle physics and the measurement of information. Unfamiliar though it is, Shannon's formula nevertheless has imme-diate practical consequences. It implies that you can improve communications in two ways. On the one hand, you could increase the frequency, measured in pulses per second, of the signal. This is the reason for the substitution of light waves in optical fibres for electrical signals in copper wires: the frequency of light is a million times greater than that of the signals scurrying through your radio and TV set. On the other hand, you could also increase the signal-to-noise ratio, but that's a much tougher task. Only the log, not the actual improvement in signal strength or in suppression of noise, contributes to the effort. You have to go way out on the x axis of the log function to register a significant increase in its height. For those of us who, by reason of sickness or age, suffer from hearing loss, the consequences are as familiar as they are unpleasant. A reduction by half in the frequencies we perceive can be overcome by a concomitant increase in the signal-to-noise ratio of the message we are interested in – but not by a mere factor of two! The log in Shannon's formula has the nasty implication that the new signal to noise ratio must be *squared* to restore the transmission of information to its former value. If S/N was, say, five before the loss of hearing, it must become five squared, or twenty-five, not just ten, to overcome the handicap of a 50 per cent hearing loss. That's why my family won't let me watch TV when they're in the room – the volume I need is too high for them. And that's why cocktail parties, with their drastically heightened noise level, are hopeless for me – at least for purposes of information transfer.

Yet noise has been unjustly maligned. Some noise is so desirable that it is introduced for the explicit purpose of masking other, more unpleasant sounds. Fancy electronics catalogues advertise generators of pleasant surf and forest sounds, as well as something more generic called 'white noise', for sensitive sleepers. A different use of noise is even more beneficial. Jan Kåhre's book on information, to which I will return later, invites us to contemplate the sailor tapping the barometer he is about to consult. The disturbance he thus introduces counteracts friction between the mercury and the glass, and leads to improved readings. Other examples include a singer's vibrato, which turns out to be necessary for staying on pitch, the high frequency tremor in the human eye that seems to play an essential role in vision, and the chopping of steady electrical and optical signals to produce alternating currents that are more easily amplified than direct currents. (In some of these cases the word 'noise' may be inappropriate for the extraneous disturbing signals that are purposely added to the original messages in order to enhance control.)

An even more dramatic instance of the usefulness of noise is a phenomenon called 'stochastic resonance', which has recently been found to play a role in a great variety of settings ranging from biology to quantum physics. The idea is simpler than its fancy name suggests. Consider a weak signal, such as a desperate plea for help tapped out in Morse code by a sailboat in trouble – with the battery failing. Imagine further that the radio receiver of a nearby freighter has a minimum threshold of detection, so if the incoming signal is below a certain strength it will not register. But now imagine that the sailboat's equipment suffers further degradation, and starts to emit a useless buzz in addition to the SOS message.

If that additional noise is enough to boost the outgoing signal above the threshold of the freighter's receiver, an alert radio officer just might be able to make out the faint distress signal poking its head over the edge of what he can detect. Normally the noisy hiss of the dying radio would drown the coded signal, but the peculiarity of the receiving apparatus, with its cut-off limit, saves the day. The strength of the noise must be just right for this trick to work. If it is too low, it won't boost the signal over the threshold. On the other hand, if it is too loud, it drowns the signal altogether.

Stochastic resonance is a delicate mechanism, but astonishingly effective when it works. The word 'stochastic' derives from the Greek for 'aiming', and means 'scattershot' or 'random'. In this instance it refers to the random noise drowning the signal. Resonance, on the other hand, implies control and fine tuning for the purpose of enhancing a feeble effect. For example, when a wine glass happens to resonate with the pitch of a singer's voice, it shatters. In stochastic resonance, however, the collaboration is not between two well-defined frequencies, such as the singer's pitch and the vibration frequency of the glass, but instead between a regular signal and a random background, between information and noise. Such cooperation seemed so unlikely that it remained undiscovered until 1981, when it surfaced unexpectedly in connection with an obscure problem in climatology. Today stochastic resonance is invoked to help explain how animals and humans can see and hear faint signals in noisy environments, and possibly even how our brains work. Traditionally the enemy of information, noise is becoming its partner.

The most important role of noise, however, is as the preserver of our sanity. Without noise, the measurement or observation of a single physical quantity would require an infinite memory and an infinite amount of time – it would overload all our circuits. Neither science nor consciousness could exist. If the world is thought of as an infinitely complex and sharply detailed landscape in which we dwell, then noise is a thick blanket of snow that softens the contours into large, rounded mounds we can perceive and sort out without being overwhelmed. Time has been called God's way of making sure that everything doesn't happen at once. In the same spirit, noise is Nature's way of making sure that we don't find out everything that happens. Noise, in short, is the protector of information.

15 Ultimate Speed

The information speed limit

Send your friends a letter and it might be with them the next day. Send them an e-mail and it will reach them in the blink of an eye – which leads to an interesting question: how fast can information travel? Clearly, the answer must have weighty commercial consequences, but it also happens to throw light on the concept of information in a way that renders it even more revealing from a fundamental point of view.

A first guess might be that, since information resides partly in the mind, it moves, as Samuel Morse put it, at the speed of thought. As an example, imagine that you and I have two marbles, one blue and one red. We each stick one into our pocket without checking its colour. While you remain at home, I strap myself into my rocket ship and zoom off to the other side of the Milky Way galaxy. On arrival, I have no idea which marble you have in your pocket; but if I pull mine out to look at it, I will immediately know the colour of your marble a zillion miles away. Information seems to have travelled across the galaxy in an instant. But it hasn't. The information I acquired was hiding in my pocket all along. After travelling from my hand to my eyes in a twinkle, it entered my brain, and encountered my prior knowledge – the correlation between our two marbles. The rest was logic, not communication across a vast distance.

Could a variation of this scenario be used for instant transfer of information? Suppose that, after I look at my marble, I want to inform you back on Earth that I am now aware of the colour of your marble. The only way to do that would be to send a radio signal, or a laser pulse, or a letter in a rocket, back home to you; and then I would run into two huge stumbling blocks: that awesome distance, and Einstein's theory of relativity.

Information is not a thing like a table, an atom or a light beam; but to carry information from place to place it must be encoded on something, on a vehicle of some sort, and all the vehicles we know of are, in turn, covered by relativity. The theory of relativity distinguishes two fundamentally different classes of objects that might be used to carry information. On the one hand there are photons, which constitute light, radio signals, X-rays, and all other electromagnetic waves. Photons, and their siblings in the elementary particle zoo, never slow down in a vacuum, or stop to be weighed, so their 'rest mass' – the mass they would have if they could come to rest – is set equal to zero. For this reason they are loosely called 'massless', even though in flight they do carry mass/energy. In modern terms, relativity declares: *Massless particles always travel through the vacuum at the speed of light* (universally abbreviated c, for celerity). This rule follows from the laws of electromagnetism, which imply that its waves always travel through a vacuum with speed c, nothing more, nothing less. The second class of objects comprises massive particles like electrons, atoms, molecules and baseballs. Einstein discovered that massive particles gain weight when they speed up, and simply grow too heavy to accelerate further as they approach c. So massive particles, unlike photons, *can come to rest, but cannot reach c.*

Contrary to some accounts of special relativity, Einstein never used words like 'information', 'signal', or 'message' when he formulated the theory in 1905. Only the vehicles that carry them are mentioned. This raises the interesting possibility that there might be other vehicles – neither massless nor massive particles – that can be used to ferry information, and that might beat the speed limit c.

Let us suppose that out there somewhere such vehicles exist. Specifically, imagine that a new type of radiation is discovered and clocked at four times the speed of light. I'll call its rays 'iris waves' after the goddess Iris who flits through the *Iliad* delivering messages. Let us further assume that the theory of relativity is correct, and that it covers iris waves along with the rest of the material world. Einstein realized shortly after devising his theory that his rules for computing speeds allow you to send iris waves, or their equivalent, *backwards in time*. In this way you would be able to transmit messages to your own past, have your mother killed before you were born, and generally play havoc with cause and effect. Since this scenario is absurd, we are forced to choose between belief in iris waves and acceptance of the theory of relativity. As long as there is no evidence to support the former, we follow Einstein's advice and adopt the latter; but there is no proof that iris waves won't ever be discovered!

Faster-than-light communication is incompatible with special relativity, but it nevertheless haunts the professional and popular literature with dogged persistence. The issue first came up two years after the introduction of relativity, and even now hardly a year goes by without some new report of superluminal signalling. One reason for this is that, although no one disputes the analysis of how things travel through a vacuum, strange things happen when light shines through a material medium like air or glass. In certain circumstances, speeds greater than c show up. Unfortunately it is not so simple to figure out just what it is that is moving at that rate. If it turns out to be a real thing, like an iris wave, relativity will be in trouble, but if it is as insubstantial as your knowledge of the colour of your marble, it won't amount to more than a curiosity and grist for the science-fiction mill.

At the annual meeting of German Physicians and Scientists, held in Dresden in the autumn of 1907, at a time when the theory of relativity had not yet entered the mainstream of theoretical physics, the great theoretician Arnold Sommerfeld, who was to become mentor to a generation of quantum physicists, gave a short lecture with the intriguing title 'Disposal of an Objection to the Theory of Relativity'. Sommerfeld pointed out that when a transparent medium happens to absorb a certain colour of light, a signal ray with a similar colour is not only partially diminished by absorption, but also badly distorted from its original shape. This happens, for example, when blue light is sent through a filter that absorbs most of it. If the signal enters the medium as a smooth, sinuous wave, it will emerge transformed into a complicated wiggle. Different parts of that wiggle travel through the medium at different speeds, and well-established optical theory predicts certain speeds associated with them that exceed c. Sommerfeld, who was one of the world's first physicists to appreciate the full significance of Einstein's revolutionary discoveries, took this objection seriously.

Material media that split a wave signal into different components travelling at different speeds are not uncommon in nature. For example, if you throw a pebble into a pond, and watch carefully, you will find the spreading ring of water corrugated by other smaller ripples that seem to come from its outside and disappear into the calmer surface near the centre. Both are waves moving at different speeds, which have acquired the names 'phase velocity' and 'group velocity', respectively. Sommerfeld showed that the optical analogues of both these velocities may exceed c, but that neither represents the actual speed of the signal. The superluminal velocities are illusions, as it were, and are not associated with the transport of energy or of information.

A simple analogy of Sommerfeld's discovery crops up in old films. Consider a film of a wagon wheel with sixteen spokes. Initially, as the wagon accelerates, its spokes will seem to accelerate, but as the wagon continues to pick up speed, something odd may happen: the spokes may stop and begin turning in the wrong direction. (A similar effect is often seen in films of aeroplane propellers.) The illusion results from the interplay between the speed of the wheel and the repetition rate of the separate stills that make up the film. As long as the wheel turns very slowly, its spokes will advance in successive frames; but if each frame happens to show the wheel advanced by exactly one sixteenth of a revolution, each spoke will be replaced by its successor, and the wheel will appear to stand still. When the wagon rolls a little faster, each frame might catch a spoke just an instant before it has advanced by a full sixteenth of a revolution, with the result that the wheel seems to spin backwards.

Notice that the apparent speed of the wheel, which can be extremely fast, and even retrograde, is not real. It is an artefact of the subtle interplay between the real speeds of the wagon, the camera, and, occasionally, the movie projector as well. The analogy with Sommerfeld's problem is not perfect, but the story does illustrate that in periodic phenomena, such as turning wagon wheels and sinuous waves of light, spurious velocities can show up that may be smaller or larger than any actual velocities, and can even become negative.

In addition to the group and phase velocities, which may exceed c, Sommerfeld defined another speed he called 'signal velocity' to reflect realistically how fast a pulse of light traverses a medium. At the same time he introduced a powerful new mathematical technique for dealing with this problem. The line of investigation he initiated bore fruit all the way into the Second World War in connection with radio and radar waves in the atmosphere. The final verdict was that the signal velocity may reach but never exceed c, so that classical electromagnetic radiation in material media poses no threat to the theory of relativity.

Since then, experiments with individual photons have led to the new science of quantum optics, and a whole new problem has arisen. In quantum mechanics, widely separated systems are intertwined in ways that are subtler than the correlations of marbles in our pockets. Can this entanglement be exploited for superluminal signalling?

In April of 2001 three physicists at the NEC Research Institute in Princeton, NJ, published a paper that included the following sentence: '... a *negative transit time* is experimentally observed resulting in the peak of the incident light pulse to exit the medium even before it entered it.' Magic? Supernatural powers? Laser telepathy? Violation of causality? None of the above.

The vital clue is the word 'peak'. In the above experiment, our three physicists sent a teepee-shaped pulse of light through a glass vessel about six centimetres in length – a passage resembling that of a mile-long train passing through a hundred-foot tunnel. As expected, when the vessel was empty, the signal zipped through with speed c, each part of the pulse – leading edge, peak and trailing edge – taking two-tenths of a nanosecond to cross the container. When the vessel was filled with the dilute vapour of the metal cesium, however, the signal became distorted by its passage through matter. The very front edge of the pulse still travelled at speed c, by skimming through the empty space between atoms, as it were. The rest of the teepee, on the other hand, travelled more slowly, but was pulled forward a little, so that it seemed to get through faster. In particular, the peak itself arrived 63 nanoseconds *earlier* than its counterpart in the empty vessel.

The resolution of the puzzle is an elaboration of Sommerfeld's original analysis. A teepee-shaped pulse of light, or any other signal, can be thought of as a sum of a large number of infinitely long, purely sinusoidal waves. It turns out that in cesium gas, these waves travel with different speeds. Thus it can happen that if they are originally in step at the location of the peak, where they interfere constructively, they get out of synch as they travel through the vessel. Under very special circumstances they may end up in step a little bit *ahead* of where the peak was before. If you think of the pulse as a solid unit, it travels along in the shape of a teepee until its leading edge hits the cesium. The shock of the collision affects the entire signal, and causes the teepee to lean forward, so it arrives earlier. The NEC scientists were sufficiently aware of the sensational nature of their claim that they took pains to insist right up front: 'This counterintuitive effect is the direct result of the ... wave nature of light and is not at odds with either causality or Einstein's theory of special relativity.'

In a later publication, they joined forces with other researchers to look more thoroughly for the physical mechanism that limits the speed of information to c. Whereas they had originally thought that classical physics alone resolved the paradox, they now realized that quantum mechanics was at work after all. The innate fuzziness of nature introduces extraneous noise in the detectors, and blurs the time measurements sufficiently to sweep the sharp contradictions of classical wave theory under the rug. That quantum theory should come to the rescue of causality in this fashion is a reminder of the fact that light – which always moves with speed c and is consequently intrinsically relativistic, also comes in the form of quantum-mechanical bundles, and is furthermore detectable as a macroscopic wave phenomenon – is uniquely capable of illuminating the consilience of three fundamental theories: relativity, quantum theory, and classical physics.

The problem of defining exactly what is meant by the signal velocity, which cropped up as long ago as 1907, has not been solved. After engaging classical physicists for fifty years, and quantum mechanics for another fifty, it is poised to enter its second century. The second paper by the NEC physicists ends with the words: '... it can be hoped that the recent interest in quantum information theory might lead to a generally accepted notion of the signal velocity of light pulses.' For the moment it appears that information is stuck with the universal speed limit, but the search for understanding just why and how may bring new surprises. The rules may change, with unforeseen results, when distances and times are reduced downward from centimetres and nanoseconds to the atomic dimensions of nanometres and femtoseconds. In the meantime, however, the limits on the growth of computing prowess imposed by the information speed limit are but a dim warning signal of difficulties far ahead on the horizon.

16 Unpacking Information

The computer in the service of physics

In May 1952, three years before his death, Albert Einstein wrote a letter to explain the nature of science to his old friend Maurice Solovine. In it he included a little sketch that put the matter more clearly and convincingly than any fat monograph; with half a dozen strokes of the pen Einstein lays out his view, polished by a lifetime of experience and reflection, of how scientific theories are developed and verified.

At the bottom of the drawing a straight, horizontal line represents the solid ground from which all science grows, and to which it must always return. It is labelled 'E: the variety of immediate sense Experiences', and includes everything that our senses perceive, nakedly or with the help of instruments: rainbows, snowflakes, stars, symphonies and quarks. The plane of experience, as Democritus realized more than two thousand years ago, anchors science to the material world and distinguishes it from speculative philosophy.

High above that plane a dot marked 'A' means 'the system of Axioms' and stands for the great fundamental laws of science: the atomic doctrine of Democritus, Maxwell's equations of electromagnetism, the two laws of thermodynamics, Darwin's theory of evolution, Mendel's laws of heredity, and a handful of others. On the left of the drawing is Einstein's own contribution to the philosophy of science: an exuberant, swooping arrow that starts just above the plane of experience and curves around to point at A. It represents the inductive leap from observation and experi-

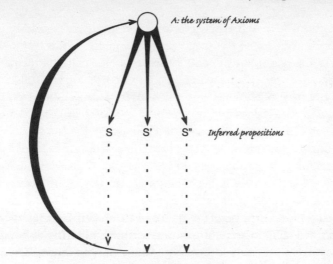

A: the system of Axioms

S S' S" *Inferred propositions*

E: the variety of immediate sense experiences

ment to theory – not, as is commonly assumed, a logical or methodical inference, but a 'free invention of the human intellect', as Einstein put it elsewhere. Feynman called it a guess. The inductive leap connotes inspiration, imagination, invention, intuition, insight and instinct. The inductive arrow does not spring up from any one point in the plane E, but skims along over it for a little distance, gathering evidence, as it were, without requiring a firm attachment to specific facts.

On the right of the drawing, three straight arrows point down from A to three points labelled S, S', and S" located halfway between E and A. They stand for 'inferred propositions', those rigorous mathematical deductions from the fundamental laws, such as the proof of Kepler's laws of planetary motion from Newton's law of gravity. If the inductive leap is the work of the great creative geniuses of science, the deduced propositions are produced by ordinary theoreticians.

Finally, S, S' and S" are connected by dotted arrows to specific points back down on the plane E. The dotted lines represent the work of experimentalists, whose job is to compare theoretical predictions with actual observations, for, ideally, a theory should lead not just back to the sparse set of phenomena that inspired it in the first place, but to many others as well. Such unanticipated points on E are called predictions, and occupy a crucial place in every outline of the scientific method. In addition, those dotted arrows also serve to imply the principle of parsimony, which Einstein prized highly in a scientific theory, by the fact that while just one arrow leads up to A there are three, standing in for a much greater number, coming back down. 'The grand aim of all science is to cover the greatest number of empirical facts by logical deduction from the smallest number of hypotheses or axioms,' he maintained. The same principle is at work in Ockham's razor, in Feynman's panegyric upon the atomic doctrine and in the technique of data-compression in information technology – all three of which extol economy of expression, albeit for different reasons.

With its arched tether on the left and multiple rigid struts on the right, the point A looks firmly attached. It is this dual support, representing the interplay of induction and deduction, that lends robustness to the axioms of modern science. Neither one by itself would suffice to sustain the scientific process, but together they form a structure of unparalleled power in the history of human thought. Both induction and deduction, reasoning from the particular to the general, and back again from the universal to the specific, form the essence of scientific thinking. The German poet Goethe, using different terms, thought that among scientists 'analysis and synthesis should alternate as naturally as breathing in and breathing out'. Einstein's drawing refines that idea by exhibiting it as a continuous, clockwise cycle from factual evidence up to theoretical construction, and back down again by an easier route.

Whereas the establishment of scientific laws is a drastic act of information–compression, the deduction of a theory's consequences is just the opposite – an act akin to unpacking a suitcase bursting at the seams with information. As Feynman pointed out, it takes imagination and thinking to accomplish this unpacking, but it also takes hard work. The history of Einstein's own theory of gravity illustrates just how much.

After a struggle lasting many years and as many false starts, Einstein published his definitive paper on general relativity in 1916, which culminates in a single, highly abstract and remarkably compact formula, called the relativistic field equation, or simply the *Einstein equation*. Its notation is so terse that the first fifty pages of the original paper were devoted to fleshing out the meaning of its symbols. When it is written in terms of the more familiar coordinates that define a position and an instant in space-time, it explodes into sixteen interlinked equations, each of which may have up to several hundred entries, depending on circumstances. Since it holds at every point in space-time separately, it incorporates an infinity of relationships. The symbols on one side contain a complete description of the geometry of space-time, and those on the other, the distribution of the sources of gravity, such as the masses and locations of stars and planets. As John Wheeler put it: 'Matter tells space-time how to curve, and space-time tells matter how to move.' If this aphorism conjures up a vision of space-time as a swirling mountain stream carrying sand, pebbles and boulders past the cliffs and rocks that determine its course, you're on the right track.

The amount of information in this one equation is so overwhelming that nobody can make use of it in its full generality – indeed, even Einstein himself could only solve it in the most special of circumstances, and then only approximately. Nevertheless, no sooner was the equation published than the German astronomer and physicist Karl Schwarzschild, while serving as a volunteer on the Russian front in the First World War, and dying from a rare disease, succeeded in working out a complete solution for one idealized but important case. He began by assuming the Sun to be a perfectly smooth, non-rotating sphere, and then used this assumption to calculate the shape of space-time around it. From this he derived the force of gravity on any object in the Sun's (or any other star's) vicinity, and in doing so corrected Isaac Newton's 250-year-old law of universal gravitation. The observable effects of the correction on planets and light beams, which Einstein had only been able to estimate, could now be predicted in full detail –

and, to date, these predictions have agreed perfectly with all precision measurements.

In the decades that followed, astronomers, physicists and mathematicians found an additional handful of solutions to Einstein's
equation; but theirs was an arduous and thankless task, because so
very little of what they discovered was amenable to experimental
verification. Add to this the fact that astrophysicists were not
particularly interested in academic debates about the mathematical
properties of Einstein's equations under unusual but unphysical
conditions, and that mathematicians rarely get interested in problems whose subtlety is hidden under mountains of sheer bulk, and
it is perhaps understandable that some of the impetus was lost.
Indeed, it was not until the last decade of the twentieth century
that two major developments coincided to thrust the Einstein
equation back into the forefront of scientific research.

One of the predictions of general relativity is the existence of
gravitational radiation, a ripple in space-time analogous to the
vibrations in the electromagnetic field that we experience as light.
These gravitational waves are believed to criss-cross the universe
at the speed of light, but on account of their feebleness they have
never been detected directly. Convincing indirect evidence for
their existence (such as the loss of energy and consequent spiralling
inward of a pair of neutron stars that occur in precise quantitative
agreement with Einstein's equation), has been found, but the lack
of direct empirical corroboration of a prediction of such obvious
fundamental importance has remained a deep source of frustration
to physicists and astronomers for many decades.

Today, though, this situation looks set to change. Improvements
in laser technology, computerized instrumentation, and fast electronics have led to the construction of several laboratories throughout the world engaged in a race to detect gravity waves. Unlike
seismologists, who suspend large, stationary weights from delicate
threads and watch the ground jiggle back and forth under them,
the gravitational wave-hunters watch for the jiggling of the weights
themselves as they respond to the passage of ripples in space-time.
Teasing out the minuscule signal from a cacophony of extraneous

noises is tedious, painstaking work, but the vision of a Nobel prize for the winner sustains it.

The search for gravity waves has also inspired theoreticians to mount a fresh attack on the Einstein equation. Their immediate aim is to predict the shape of the signals that will be recorded in the laboratory. Compared to earlier attempts, the input assumptions now have to be much more realistic. (There is an old joke about a theoretical physicist hired to solve a structural problem in a dairy, who starts off with the words: 'Consider a spherical cow.' A system of colliding black holes is as far removed from Schwarzschild's star as a cow is from a sphere.) The Einstein equation is thereby rendered completely intractable to solution by hand, or even by conventional computer. Fortunately, however, super-computing has come to the rescue just when the experimental evidence is about to appear. If all goes as planned, the twenty-first century will be the era of gravitational astronomy, opening a new view of the cosmos to complement those afforded by optical, infrared, radio, X-ray, gamma-ray and neutrino telescopes.

The program of computational general relativity is as simple to state as it is difficult to carry out. You start by imagining a possible scenario of a cosmic catastrophe, such as a supernova explosion or the collision of two black holes, and solve the Einstein equation to find out how the resulting turmoil in space-time would propagate and appear to an Earth-based observer. If you then make a catalogue of the typical shapes of gravitational waves generated by different events, the experimentalists will be able to compare their finding with your calculations, and eventually match observation to prediction. It is simply a matter of constructing a new arrow from A to S on Einstein's drawing by unpacking the information encoded in his equation. The dotted arrow that is currently under construction downward from S is sure to touch down on the plane of experience some time in the coming decades.

The immensity of the computational task of deriving the shape of gravitational waves from the Einstein equation cannot be conveyed by reference to astronomical numbers of bits of data; it is more appropriately measured in human terms. Consider the career of Ed Seidel, who earned his bachelor's degree in my department at the College of William and Mary twenty years ago, studied astrophysics at Penn and Yale, and today directs the numerical relativity group at the Albert Einstein Institute for Gravitational Physics in Berlin. His curriculum vitae illustrates both the complexity of the Einstein equation and changing landscape of scientific research in the information age.

Soon after receiving his PhD, Seidel took up the study of colliding black holes and realized that the traditional methods of mathematical physics were not up to the task. Numerical techniques would have to be developed, and computers would have to be programmed. In pursuit of this goal he quickly progressed from desktops to mainframes to supercomputers, and then found that he had to aim even higher. Seidel's group of about two dozen youthful collaborators, assembled in Germany from all over the globe, includes physicists, applied mathematicians and computer scientists. This dedicated little band does not work in isolation, however, but participates in what astronomers would call 'super-clusters', associations of groups at different universities and laboratories throughout the world joined together in pursuit of a common goal. And that's not all: the collaboration extends beyond people to their supercomputers which are interconnected into 'grids' by high-capacity communication channels that can transmit the contents of a CD in seconds. An intermediate goal of Seidel's group, preliminary to the next round of computations, is to develop an intelligent network of supercomputers that can automatically scope out the availability of idle machine time on its widely scattered sites, and use it to advantage. Eventually it should even learn to subdivide the task at hand, and distribute the pieces for simultaneous processing by the participating computers. General relativity has come a long way since 1916. If Einstein was a sorcerer, and Schwarzschild his apprentice, today's computational relativists with their arsenal of computers form a proliferating army of brooms marching into battle to tame a seething ocean of data.

The scientific output of all this technological firepower is as often graphical as numerical. Physicists have learned that images, both still and moving, are efficient ways of displaying huge amounts of numerical data. Seidel's colour-coded film stills of gravitational waves emanating from colliding black holes and other astronomical disasters are not only revealing to the expert, but also beautiful to the layman. With their swirling patterns in lush colours they have acquired a new function in the reward system of the scientific community. Where the badges of success of theoretical physicists of my generation were citations of papers in the premier research journals of the world, the younger generation struts its stuff in visual terms. Seidel's website, for example, is decorated with a gallery of stunning pictures reproduced on the covers of scientific journals, all illustrating solutions of the Einstein equation. No general could be prouder of the colour-coded ribbons on his chest than Seidel is of these trophies from his digital campaigns.

Solving the Einstein equation in order to predict gravity-wave signals on Earth is reminiscent of Feynman's claim that 'the *atomic hypothesis* ... contains an *enormous* amount of information about the world'. But where Feynman's example is qualitative, Seidel's is eminently quantitative, and amenable to bit-counting. The obvious starting point for such a measurement is the Einstein equation. When it is parsed as a message, it is very brief indeed. Even written out in its full four-dimensional glory, the computer program that reproduces it is only a few hundred bits long, so its complexity is not an appropriate measure of the amount of information involved. On the other hand, taking into account the fact that the equation holds for every point in space-time, its information content jumps to infinity. Neither estimate is very useful.

Alternatively we could look at *solutions* of the equation instead, and count pixels in one of Seidel's images. Now the information content appears large and finite. Alas, it still suffers from the fundamental flaw of bit-counting: lack of meaning. A Jackson Pollock jumble of the same size contains the same number of pixels.

In search of escapes from this dilemma, a number of alternative measures of information have been proposed. The Nobel laureate Murray Gell-Mann, for example, favours a concept called 'logical depth', which deals with the amount of organization in a data set by clocking the computer time used up in its creation. Unlike 'algorithmic complexity', which examines the length of a program, logical depth looks instead at the number of steps executed by the program. The physicist Jim Crutchfield promotes a different notion called 'statistical complexity', which attempts to quantify the amount of structure in a system by measuring the size of the computer memory required to predict the next step in the system's evolution. If the system is very orderly, its statistical complexity is low because the pattern is obvious; if it is random, not much memory is required to ascertain this fact. In between, where there is a lot of interesting structure, statistical complexity is correspondingly high.

By any measure – length of programs, computer time used up, size of memory – Seidel's solutions are massive. They are rich in organization, pattern and structure, terms that we encountered earlier as synonyms of form. They are replete with information. In some way – no one is quite sure how – all this information (except for the initial conditions) is contained in the equation that originated and matured in Einstein's mind. Unpacking it with the aid of thinking machines for the purpose of comparing it to observations is what modern computational physics is all about.

17 Bioinformatics

Biology meets information technology

Shortly after the human genome was unravelled, biologists made an astonishing discovery. To their very great surprise, they found it contained a far smaller number of genes than they had expected – not many more, in fact, than go to make up the earthworm genome. Consternation ensued. Are we as simple as worms? Or are worms as complex as we are? The answer to both questions is no. The message of the genome will be found to reside not simply in the sequence of genes, but in their manifold interactions and relationships as well. To put it one way, a collection of Shakespeare's plays is richer than a phone book that uses the same number of letters; to put it another, the essence of information lies in relationships among bits, not their sheer number.

The human DNA molecule is shaped like a twisted ladder, with rungs made up of the four chemical bases adenine, thymine, cytosine and guanine, which are universally abbreviated A, T, C and G. Two immensely long chain molecules – the vertical side-pieces of the ladder – wind around each other in helical fashion. The base molecules sticking out from the two chains point towards each other pairwise. When the pairs hook up by means of weak chemical bonds, they form the crosspieces, or rungs, of the DNA ladder. The entire ladder, which is divided into twenty-four separate, unequal lengths corresponding to as many different chromosomes, comprises about three billion rungs. Reading along the length of one side of the ladder, we might encounter a string like this: TTTTCATTAGTTGGAGA . . . and so on and on and on.

The letters form 'words', called 'codons', each only three bases long, which represent one of the twenty primitive building blocks of the body called amino acids; but three bases don't make up an amino acid – they only encode it. To take a specific example, the second codon in the above string, TCA, is a cipher for serine, and the starting point for a complex biochemical minuet called gene expression, which results in the actual manufacture of serine from the raw materials stored in the human body. The next step in the process is the assembly, following the DNA construction manual, of amino acids into proteins, the molecules of living matter.

From the point of view of information-processing, the human genome is a tour de force. It's as though nature said to us: 'OK, so you have invented computers and computer science – you think you are so clever. Here's a problem to test your mettle. See if you can crack this one!' Indeed, if the genome's structure were two- or three-dimensional, molecular genetics would be well-nigh unimaginable. As it is, the genome's linear, one-dimensional structure makes it sufficiently similar to the binary code that transcription into computer language becomes trivial. The length and complexity of the entire genome seems designed to fit modern information technology hand in glove. If it were substantially shorter, it could not encode enough information to make a living organism. If it were much longer, on the other hand – say 10^{20} letters long – it would not be accessible to present-day computers.

The most remarkable property of the genome is its meaning. Its message appeals to our emotions; it *matters* in the most profound sense possible. The genome speaks neither of mere bricks and mortar, nor of complex circuits made of wires and silicon, nor of cunning games of the mind, but of the material basis of birth and health and disease and death. The genome tells the story of life itself.

In any discussion of the application of information technology to biology, called bioinformatics, it is important to keep in mind the fact that molecules are invisible. A sequence such as TTTTCAT-TAGT cannot simply be read off, even through a microscope. It represents the culmination of a painstaking process of detective work involving wet chemistry, X-ray crystallography, logic, mathematics and a vast store of knowledge gathered during the past century from experiments *in vitro* and *in vivo*. An army of scientists in laboratories throughout the world is working out the details of this enormous puzzle. The ideas that tie it all together are in the end biological, not mathematical or technological; but without information technology, genetics could progress no more than numerical relativity with its spiderweb of intimately connected mathematical equations.

The first chore of a computer let loose on the genome is rather mundane. It turns out that 97 per cent of the letters in DNA don't represent anything at all. They are junk, garbage and nonsense – at least at present. As we learn more, this assessment will surely be amended. In the meantime, though, sifting through the world's genetic databanks for clues as to what's useful and what not is clearly a job suitable for automation. In fact, even before this clean-up job, information technology faces preliminary challenges. The libraries of genetic information scattered throughout the globe use different standards, personalized abbreviations, incompatible software programs for storage and communication, idiosyncratic systems of cross-reference and undocumented subroutines. In short, biological data-retrieval is a nightmare.

Once the wheat is separated from the chaff, and reasonably standardized notations and methodologies have been agreed upon, the real work begins. At every level of genetics, both before and after the physical assembly of proteins, variability and exceptions are sprinkled through the grand scheme. The amino acid serine, for example, is indeed associated with the codon TCA, but also with the five other codons, TCT, TCC, TCG, AGT and AGC. Just as the analysis of different synonyms and different dialects helps to distinguish English texts from different sources, the occurrence of different codons for the same chemical, or of different proportions of these codons, can harbour telling biological clues.

As an example of applied bioinformatics, consider the production of insulin for treating diabetes. A certain molecule involved in this process in humans consists of a chain of about fifty amino acids. If you abbreviate each one by a letter, you get a fifty-letter 'word'. Pigs, rabbits and cows, it turns out, also produce insulin, each represented by a word. If the four words, written below each other, are compared letter by letter, a remarkable similarity shows up. The first thirty or forty letters are identical in all four examples, but then slight differences show up. In the end, the overlap between human and animal insulin, measured by the number of matching letters divided by total length, is around 90 per cent. Whether these differences are serious enough to cause trouble if animal insulin were used on humans, or whether they are examples of nature's redundancy and do no harm, are questions for biology and medicine. When examples like this are multiplied to range over the entire genome, the full extent of the animal kingdom, and all the diseases and anomalies found in nature, it becomes apparent that computers are not a luxury but a necessity.

The most difficult job facing bioinfomatics brings the subject of information right back to its original roots. The footsoldiers of human physiology, who perform the actions that make a cell grow, thrive and function, are the proteins. They come in countless varieties specifically adapted for as many specific functions. They are built up from linear molecules copied according to the DNA blueprint, but they are not linear themselves. In fact, they are usually kind of round and lumpy, like potatoes. This miraculous transformation from one to three dimensions is performed in a manoeuvre called protein-folding. For a biologist, understanding protein-folding is up there with those other big challenges: the brain and the enigma of consciousness.

The basic idea is to be able to predict how a chain-like molecule, whose assembly from basic building blocks is assumed to be perfectly understood, folds and wads itself up into a three-dimensional structure. The process can happen in a matter of seconds, or it may take hours. Each link in the chain is firmly attached to the preceding and following links, but there are additional weak forces – reminiscent of the stickiness of a strand of overcooked spaghetti – that hold the molecule together after it is wadded up. This much is not hard to understand, and the bunching-up of a strand of spaghetti into a little ball is not a bad image to keep in mind. But a protein molecule has an added marvellous property. It has a definite, fixed configuration that defines its function, and every time it is copied, it comes out the same. It's as if a child balled up every strand of spaghetti in a pot, and each ball turned out to be identical in shape, down to the last twist and turn.

Here genetic spelling errors, inherited or introduced by mutation, can have devastating consequences. The change of a single letter in a long chain can change the code from one amino acid to another, and lead to distinctive changes in shape of the resulting protein. A tragic example of this sort of error is sickle-cell anaemia, the first disease definitely associated with a molecular origin. It is caused by the replacement of the amino acid glutamine by valine, resulting in the malformation of a single protein out of about 300 that make up haemoglobin. On the surface of the misfolded protein a bump appears that fits neatly into a pocket on an adjacent protein. Consequently the two proteins clump together, and initiate a chain reaction in which long fibres of malformed haemoglobin molecules are eventually formed, leading to a blood disease that is usually fatal.

From the medical point of view, the significance of protein-folding lies more in the exception than in the rule. In a process of this complexity, mistakes are bound to occur, and misfolded proteins perform their functions poorly, or not at all. Many diseases in addition to sickle-cell anaemia, including Alzheimer's and haemophilia, have been traced back to misfolding. Understanding how a molecule should fold properly, and why it didn't, are obviously steps on the way to figuring out how to prevent or correct errors in the process.

From the point of view of physics, the problem is devilishly difficult. At one end of the scale, we are used to asking how three or four atoms can come together to form a molecule, and at the other, how billions of them can arrange themselves into some kind of regular crystalline order. We even know something about how a snowflake grows, and how DNA is formed; but protein-folding is more difficult. Picking out trends and regularities among the infinite number of shapes into which a chain can bend itself is a problem of a new order of magnitude. Without computers it is hopeless, but even with the best computers, brute-force methods, such as trying out all possible configurations, are not viable. New computer techniques, new mathematical methods, new ways of thinking are required.

A very simple variant of the problem is no further away than your TV set. In a black-and-white set, the primary information coming from the antenna or cable is an endless stream of black and white dots. By themselves, they form a meaningless ribbon like the sequence of bases in DNA. Your monitor chops this infinite ribbon into little segments as long as your screen is wide, neatly stacks them up like lines of text in a book, and starts over again when the screen is full. In this way a secondary layer of information emerges. It is contained in the spatial relationship of the dots with others downstream, and when they are all put together in the right order – when the stream is 'folded' properly – a two-dimensional visual image, analogous to a properly folded, three-dimensional protein molecule, appears. Here too mistakes can occur, resulting in partial images, jittery pictures, or utter chaos; but since the segments that need to be lined up all have precisely the same length, and the 'folding' is as precisely regular as that of jeans in a Levi's outlet, the problem is far, far easier than its biological counterpart.

Bioinformatics starts with reams of information, stored in data banks all over the world, most of it spotty, some of it faulty, all of it complex. 'Noisy' is the word information scientists apply to this environment. The vision at the end of the process is a beautiful three-dimensional, colour-coded model of the finished protein molecule that can be held between virtual fingertips and turned this way and that, like the lacy blossom of some exotic flower.

With mountains of data, lots of errors and misinformation, and a paucity of theoretical guidance, the science of bioinformatics needs an organizing principle lest it diverge into a meaningless jumble of disconnected facts. One promising avenue of approach is the use of Bayesian methods. If every researcher were to first estimate the prior probability of the truth of some hypothesis – from the identification of a gene to proposed protein structure – and then measure progress by estimating the posterior probability on the basis of new information, the field might gradually converge to a series of well-established propositions. Whereas Bayesian thinking might play a fundamental role in a new understanding of quantum theory, it only serves a methodological function in bioinformatics. Nevertheless, it is noteworthy that the two principal achievements of twentieth-century science, quantum mechanics and genetics, are both turning to the concept of information, and the rational estimate of its value, in hopes of making progress.

The study of computational relativity described in the last chapter and the attempt to understand protein-folding proceed in opposite directions on Einstein's sketch of the scientific method. The former, as we saw, is strictly deductive. The latter, on the other hand, begins with the data resident in DNA and tries to come up with rules and regularities by induction – a task that can be made more systematic with the help of Bayesian methods. Together these two examples illustrate the two paths to scientific knowledge with uncommon clarity. Einstein was able to take both of them with paper and pencil, but in this century both the compression of data into comprehensible laws and regularities and the unpacking of axioms to obtain descriptions of observable phenomena will only be possible with the use of the computer.

With protein-folding, the term 'information' has returned to its origin. The information needed to fold a protein resides in the sequence of the amino acids that make it up; but knowing that order doesn't tell you much about folding. As the TV monitor example makes clear, structural information is encoded in relationships between amino acids, and digging those out of the primary sequence is a formidable (!) task. The genome is a code for transmitting the *form* of the molecular constituents of the human body from generation to generation, and gene expression is the act of *in-forming* – of giving form to – the atoms that serve as raw materials. The genetic code is nature's way of using information to determine structure, or form. We are beginning to come close to replicating this trick of creation, but for the moment we still resemble one-dimensional earthworms, craning our necks in wonder out into the second and third dimensions.

18 Information is Physical

The cost of forgetting

While some historical eras are defined by the personalities of their leaders, some by religious doctrine and some by great social upheavals, others, such as the Industrial Revolution, which was powered by the steam engine, and our own information age with its technology based on the computer, are dominated by their machines. Both of these devices were developed by engineers with precise practical goals in mind, and both only later attracted the attention of scientists intent on understanding their fundamental principles of operation. In the case of the steam engine, the emerging science was thermodynamics; in the case of the computer, it was information theory – disciplines that both began with a search for the limits imposed on technology by physical law, but came to remarkably different conclusions.

Thermodynamics, the science of warmth, was founded in 1823 by the French military engineer Sadi Carnot. On page six of his *Reflections on the Motive Power of Fire* we read the words that herald its birth:

> In order to consider in the most general way the principle of the production of motion by heat, it must be considered independently of any mechanism or any particular agent. It is necessary to establish principles applicable not only to steam engines but to all imaginable heat engines, whatever the working substance and whatever the method by which it is operated.

The switch from 'steam engines' to 'heat engines' signals the transition from engineering practice to theoretical science.

Sadi Carnot was born in 1796 in Paris on the periphery of the court of Napoleon, and was named after a medieval Persian poet. His father Lazare, himself a cultured man of letters, engineer, general and statesman, oversaw Sadi's broad education in music, languages, mathematics and science. Eventually Sadi studied engineering and joined the army, but at age twenty-four he took leave as lieutenant on half-pay in order to devote himself to his research. His wide interests included political economy and public education, but his main passion was what today would be called theoretical physics. A thorough survey of industrial development had convinced him of the significance of the steam engine. When he realized that its perfection over the course of half a century had come about almost exclusively at the hands of the English, he decided to rectify this imbalance. His claim to the title 'founder of thermodynamics' derives not so much from his actual contributions, which were only partially correct, as from the way he went about making them.

After its invention by James Watt, the steam engine had been improved by engineers. They had tinkered with the mechanical design, the levels of the boiler's pressure and temperature, the choice of materials for construction and countless other variables. Even so, by 1830 steam engines were no more than about 6 per cent efficient – meaning that they wasted 94 per cent of their fuel. Carnot knew that, if he was to make real progress, he had to delve deeper. Engineering had to make way for science.

Carnot's key insight was his observation that *all* heat engines seem to throw away heat. Think of coal and oil-fired power plants puffing smoke and steam into the air, automobiles emitting hot exhaust gases, or even nuclear power plants using cooling towers or nearby rivers to dump huge quantities of expensive energy into the environment, and you get the picture. Up until that point, the challenge for engineers had been to minimize such waste by clever design. Carnot turned that logic on its head by coming to the unexpected conclusion that all heat engines – which is to say all engines that use heat as an intermediate form of energy – *must* waste heat. The final limits to the efficiency of steam engines, he proved, are not set by engineering skill, but by the laws of nature.

Carnot's reasoning differs markedly from the style of proof pioneered by Isaac Newton, and adopted by most branches of physics. It relies on logic and the description of imaginary devices, rather than on deductions from axioms formulated in mathematical terms. To demonstrate the truth of his proposition, Carnot uses a 'contrapositive proof': assume the opposite and show that it leads to a contradiction. In modern terms, the argument runs somewhat like this: consider an ordinary window air conditioner. It takes heat from a cool place (the room) and transports it to a hot place (outdoors) at the cost of energy (the electric grid). Now assume, counterfactually, that there exists a machine, which we shall call an 'ideal generator', whose sole function is to generate electricity from heat, without any waste. Disconnect your air conditioner from the wall socket, and plug it instead into this device, which is powered by outdoor heat. The two machines, considered as a single unit, constitute a 'perpetual air conditioner'. Whatever energy it consumes to make electricity is delivered back to the outdoor environment in the form of heat radiating out of the back of the machine outside the window. In other words, the room gets cooler, but the perpetual air conditioner's net energy consumption is zero! *But such a thing does not exist.* Every child knows that without external help, heat runs downhill, as it were, from hot to cold, not the other way round. A cold room can warm up in the summer without the expenditure of electricity, but a hot room cannot spontaneously get cooler. Since a perpetual air conditioner does not exist, its key component, the ideal generator – a perfect heat engine – cannot exist either, and Carnot's assertion is proved. (Notice that substituting an *ordinary* generator driven by outdoor heat for the assumed *ideal* generator would upset Carnot's clever argument. It would dump its waste heat into the room, thereby defeating the purpose of the air conditioner.)

Carnot did not live to see the fruits of his genius. He died from cholera in 1832, only thirty-six years old. Twenty years later Rudolf

Clausius put thermodynamics on a firm mathematical basis. He refined and restated Carnot's proposition – that all heat engines waste heat – as the rule that entropy tends to increase, and enshrined it as the Second Law. To this, Clausius added the First Law – the conservation of energy – which had been discovered in the interim. (The misconception that heat was a form of matter called caloric, rather than a form of energy, had prevented Carnot from formulating the Second Law correctly. There is some evidence that he was on the verge of rectifying his error when he died.) With Clausius's clarifications, the world had to learn to swallow a bitter pill: heat engines are inherently inefficient. The laws of thermodynamics do not allow Detroit engineers to harness the heat and pressure of your car's exhaust for conversion to useful work, and improvement of your fuel mileage. About two-thirds of the fuel you buy is wasted, and there's nothing you can do about it, with the exception of buying an electric car. (Electric motors use no heat and are therefore exempt from the laws of thermodynamics, but you must not forget to ask how the electricity stored in its batteries is generated in the first place.)

A century later, in the 1950s, computers were developed and engineers as well as physicists began to ask about *their* limits. How big, how small, how fast, how efficient, how powerful can these machines become before they are stopped by the Second Law of Thermodynamics, the uncertainty principle of quantum mechanics, and the ultimate speed of relativity? A hundred years of experience with the laws of thermodynamics had brought home the fact that there was no free lunch. It was almost inevitable, therefore, that the conviction would arise that every step of a computation, and even every move of a message from one place to another, necessarily costs energy. To be sure, the hypothetical minimum amount of energy needed to manipulate one bit is a billion times smaller than what the real, inefficient, less-than-optimum computer on your desk actually eats up for the same task. Friction in moving parts and resistance in electrical components see to it that the minimum won't even be in sight until a couple of decades from now, but it pays to be prepared. Carnot long ago demonstrated the value of understanding waste heat in a fundamental way.

It is easy to make a plausible guess at the energetic cost of switching a bit from zero to one, or from one to zero, or to transmit it from one memory cell to another. This cost is an atom of waste, as it were, the quantum of heat created by the manipulation of an atom of information – amounting roughly to the energy of a single molecule of air skittering about in a warm room. In order to strengthen the plausibility of this estimate, scientists went on to invent and analyse imaginary little devices of various degrees of complexity and, to no one's surprise, each one could be shown to produce heat at that predicted minimum rate. By the 1960s this established lore had made its way into respected textbooks on computing and information theory.

Amazingly, all these arguments were wrong – very wrong in a very simple way. Rolf Landauer, who launched the train of thought that eventually exposed the error, called the blind acceptance by the physics community of the minimum energy dissipation 'one of the great puzzles in the sociology of science'. For, consider this: if somebody describes a device that discards an atom of waste for every bit it manipulates, and others come up with additional, unrelated gadgets that do the same thing, what has been demonstrated? Only that *some* machines behave that way, not that they all do. An argument like Carnot's, as refined by Rudolf Clausius and later thermodynamicists, is more subtle. Instead of presenting us with a list of diverse examples of engines that waste heat, it asks us to imagine one that didn't, and then cleverly leads us into a contradiction – an infinitely more powerful logical gambit.

Landauer spent most of his career at IBM as an academic physicist in an industrial setting. At mid-century his employer had begun to realize, as Marshall McLuhan put it, that the future lay in learning to manage information, not merely in producing better office machinery. In support of this mandate, IBM freed Landauer from the normal application-directed chores of industrial science and let him spend his time thinking about the effects of physics on computing at the most fundamental level. His approach is symbolized by the title of a famous article in which he summarized the state of the art in 1991: 'Information is physical'. The idea is that information invariably encodes on real, physical objects, ranging from wooden abacuses and sheets of paper to semi-conducting computer chips and neurons in the brain. Being physical, information is subject to the laws of physics. How, Landauer asked, do these laws restrict the storing, transmitting and processing of information?

In 1961 Landauer succeeded in locating the precise point in any computation in which energy is converted to heat and carried away as waste. In sharp contrast to received wisdom, he found that the manipulation and transmission of a bit of information does not necessarily entail dissipation of energy, or, as a computer designer might put it, a thermodynamic cost. His discovery has since been called *Landauer's principle*: the only step that involves an unavoidable expenditure of an atom of waste is the *destruction* of a bit of information by erasure or by the resetting of a register. For example, if a computer stores a 3 somewhere, and adds a 5 to it, these two digits will be erased and an 8 put in the place of the original 3. If an infinite memory were available, and you never had to clear a file to make room for new information, the computer could operate at zero cost; but a finite memory, or a finite magnetic tape, has to be erased before the next computation can commence. The energy wasted in that seemingly innocuous cleaning operation is the cost of forgetting.

The logical advantage Landauer enjoyed over his predecessors – reminiscent of Carnot's advantage – was that in order to refute what amounted to decades of folklore he only had to invent a single example of a lossless computation. The device on which he proposed to perform this feat didn't even have to be realistic; it just had to be possible in principle, like motion without friction, or electrical conduction without resistance. (The former is possible in a perfect vacuum, the latter in a superconductor.) Nor did the gadget have to be conventional. One celebrated design of a computer that operates without friction represented mathematical operations by the paths of billiard balls flying through a vacuum and bouncing elastically from cunningly placed paddles. Since the physical principles underlying this device are clearly understood, and frictional effects can be reduced arbitrarily without violating any laws of physics, the machine provided a particularly convincing demonstration; but who would ever dream of building such a klutzy gadget?

After explaining exactly how computers dissipate energy, Landauer proceeded to make the same error that had plagued his predecessors: he assumed. Guided by his thorough familiarity with computing machinery, he persuaded himself that the destruction of information, which happens over and over again in real computers, is a necessary feature of *all* finite computers. It took him more than a decade to be convinced by his IBM colleague Charles Bennett that this assumption, too, is false. Bennett showed that once a large but finite lossless computer has finished its task, and the result has been recorded, you can run the program backwards to undo all the steps it took previously. In this way it eats all the gobbledygook that is left over in its memory cells, and returns to the initial state it started with – without a single erasure. Thus Landauer and Bennett proved in principle – if not in practice – that the laws of thermodynamics present no limits to the growth of computers. Computation does not have to create waste heat. Computation is cool! This counterintuitive result stands in sharp contrast to Carnot's equally surprising discovery that a heat engine cannot escape the severe limit on its efficiency imposed by those same laws.

Rolf Landauer, who died at age seventy-two in 1999, was a classical physicist in an age of classical computing. He encouraged his younger colleagues to venture into the new science of quantum computing, but remained sceptical of its practical eventual usefulness. Having devoted a distinguished career to the journey along a steep, zig-zag path to understanding the principles of trafficking in bits, he left to his successors the task of starting over with qubits.

Quantum Information

19 The Quantum Gadget

Quantum weirdness brought to light

Science requires gadgets to bring the unseen within the grasp of our senses. The telescope makes astronomy possible, the microscope is essential to biology, and without the particle accelerator there could be no nuclear physics. Compared to the monumental levels of complexity and cost that such machines can attain, though, the device that opens a door to the eerie world of quantum mechanics is very primitive. Called a beam splitter, it is nothing more than a chip of glass, coated on one side with a thin, colourful layer of metallic material.

Beam splitters have been used by scientists for over a century, and have recently found their way into our homes and cars, where they help to control tiny flashes of light from the miniature lasers that lie at the heart of all CD players. Put simply, their purpose is to divide a pencil of light into two branches that take off in different directions. The most basic are lossless, meaning that they don't convert any of the light's energy into heat; symmetric, meaning that they work equally well forwards and backwards; and fifty-fifty, meaning that they split the beam into two equal portions: both of their outgoing beams are exactly half as bright as the incoming one. To get an idea of what they look like, I visited the laser lab of my colleague Bill Cooke, who uses them routinely in his work. He opened a drawer and carefully pulled out a large plastic tray that reminded me of a make-up kit. A couple of dozen glass discs, about a quarter of an inch thick, and varying in diameter from that of a penny to that of a chocolate-chip biscuit, lay neatly lined up in separate compartments. Some glistened in shiny shades of pastel, some were golden, others silvery – different-coloured coatings, Bill explained, that would reflect different colours of light. In practice, a laser beam shining obliquely upon such a surface would be partially transmitted right through the glass to the other side, and partially reflected, as by a mirror. In some cases the colour that was actually reflected depended on the angle at which the light arrived, resulting in the intriguing play of rainbow hues I saw when I slowly turned a disc while peering through it at a ceiling lamp.

It takes just two of these pretty baubles to build a simple gadget with the power to transport us down to the atomic realm where quantum mechanics replaces Newton's mechanics, and quantum weirdness rules. Indeed, those same, seemingly innocent little glass discs can reveal a lot about quantum information, too. Thus, in the spirit of reductionism, it pays to begin by taking a closer look at them – and as we do so, it is worth keeping in mind two very different ways of explaining the gross features of their operation. The first of these is the ancient corpuscular theory of light, which likens a beam of light to a volley of grapeshot, and the second is the nineteenth-century view that light consists of waves.

How, then, can a beam splitter give us these wonderful insights? To begin with, imagine using a very dim light source, such as a laser beam equipped with a variable dark filter to cut its brightness close to the point of extinction. Let it shine on a beam splitter, and have two photo-detectors (instruments far more sensitive than the human eye or a photographic plate, which are designed to pick up very low intensities of light), labelled T and R, measure the amounts of light in the transmitted and reflected beams respectively. If the detectors are equipped with an audible signal to mark the arrival of light, both will at first emit a steady buzz; but as the intensity of the source is reduced, the hum resolves itself into discrete clicks, like those of a Geiger counter. Remarkably, all the clicks are equally loud, indicating that each one announces the arrival of the same amount of energy. As the light gets dimmer, the clicks do not diminish in loudness – they just occur less often, and will continue to do so long after the beam has become invisible.

The discreteness of the clicks is the first sign of quantum physics, and something that cannot be explained in terms of waves alone. Sound, for example, dies smoothly as the agitation of the waves that carry it through the air diminish – how many times have you heard the ringing of a distant church bell degenerate into a series of discrete pops? A beam of light, on the other hand, consists of a stream of light particles, called photons, and if the light is of a single colour, each photon carries precisely the same amount of energy. As the beam dims, fewer photons arrive, but each one causes a similar click in a detector. For this reason, photo-detectors are also called photon counters.

As the drumbeat of the counters monitoring our beam splitter continues, an alert listener might soon discover a second quantum mystery: clicks from T and R occur in a random, unpredictable pattern: TRRTTRTRRRT ... On average, there will be as many Ts as Rs. If you listen closely, you will find that there are also equal numbers of pairs of the four possible types TT, RR, RT and TR, and so on for all possible groupings. In fact, the sequence is random in every mathematical sense of the word.

Besides the graininess of light, the beam splitter thus reveals the unpredictability of the quantum world. Unlike the path of a particular pellet of grapeshot, which could be calculated beforehand with some accuracy if one had sufficient knowledge about its prior motion and the design of the obstacle that splits the beam, the action of a beam splitter is absolutely, irreducibly, quantum mechanically random and unpredictable. In its rat-tat-tat you can hear the clatter of God's dice; and it turns out to be useful, too: in chapter 12, where I explained the need for random-number generators, I mentioned that such a device on the market today is built around a beam splitter.

Graininess and randomness are fundamental features of quantum behaviour; but they don't exhaust the gamut of quantum weirdness. There are other, far stranger phenomena as well. One of the most celebrated is wave-particle duality, the puzzling, counterintuitive ability of atomic systems to resemble particles under some circumstances and waves under others. Most writers illustrate this schizophrenic performance by reference to a demonstration performed two hundred years ago by Thomas Young and known as the double slit experiment. In this choice the textbooks follow Richard Feynman's lead, who saw in the double slit 'the only mystery of quantum mechanics'; but in the twenty-first century we can do better. A related experiment, using lasers, beam splitters, and photon counters in place of Young's pierced curtains and rays of sunlight, puts the mystery into modern terms and sharper relief. More importantly, as the measurable output data it produces is discrete rather than continuous, and the analysis is therefore digital rather than analogue, it can be more readily couched in the language of information.

The experiment in question makes use of a simple arrangement of two beam splitters into a quantum gadget called an interferometer. (Technically it is an interferometer of the Mach-Zehnder type. The surname of Ernst Mach, a colleague and nemesis of Ludwig Boltzmann, is familiar as a unit of supersonic speed.)

Light emerging from a beam splitter is quantum mechanical, but as soon as its photons are absorbed in detectors and announced by the loud clicks of the counters, its true nature is destroyed. Meters and detectors, which communicate atomic information to our senses, squeeze all quantum subtlety out of a system, leaving only a residue of classical information. In order to capture the delicate quantum-mechanical correlations between the two halves of the beam – while they are still in their quantum state, as it were – the two parts have to be brought back together again for comparison purposes before they are swallowed up by detectors. This is accomplished by means of two diagonal mirrors, which deflect the two divergent beams onto the opposite faces of a second, identical beam splitter.

Tracing the fate of the pencil of laser light, which splits first into two and then into four parts, and leaving out of consideration the auxiliary mirrors, reveals that the beams that were twice reflected (rr) or twice transmitted (tt) by the beam splitters emerge together

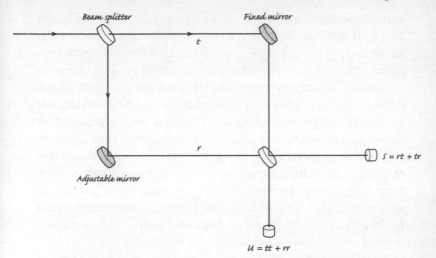

from the lower side of the second one. The other two, each of which was once transmitted and once reflected (rt and tr), come together on the upper side. The intensity of the first pair of combined rays (rr + tt) is measured by a detector called U, for 'unsymmetric', because the fates of the two beams it monitors are different, while a second meter, dubbed S for 'symmetric', measures the second pair combination (rt + tr).

Now imagine that the lengths of the paths between the beam splitters, via the mirrors, are adjusted to be precisely equal. In practice, this condition is impossible to realize, but physicists have no trouble accounting for the inevitable minute path differences that we are blithely ignoring here. If, then, the paths are made equal, the experimenter discovers a stunning surprise: the meter labelled U detects nothing at all! Instead, all the light that entered the apparatus emerges unscathed at the S meter. Even though the first beam splitter divided the laser beam into two rays, each of which was then split once again, nothing gets through to U. This unexpected result is in violent conflict with the particle theory of light. Imagine building an interferometer for a stream of grapeshot, using mechanical beam splitters. (A perforated metal plate that lets half the pellets pass through its holes but deflects the other half, and is mounted at 45 degrees to the beam, would serve the purpose.) There is no plausible mechanism to account for the failure of U to detect half of the pellets, and S the other half. Lead pellets simply do not have the ability to cancel each other out.

Waves, on the other hand, do! The trick is simple: when two identical waves happen to meet in such a way that a crest of one always encounters a trough of the other, filling it in, and a trough always meets a crest that it can swallow up, the result is indeed zero. Waves can, and under the right circumstances do, annihilate each other in a phenomenon called 'destructive interference'. Of course, the other alternative, that crest meets crest, and trough hits trough, creates a bigger wave in a process called 'constructive interference'. In our interferometer, constructive and destructive interference go hand in hand: the energy missing from meter U shows up in S.

With this realization, the failure of the U detector to catch any light loses its mystery. The two pencils of light that emerge from a beam splitter, though they have equal speed, frequency, colour and intensity, are not absolutely identical. They differ in one subtle respect that is not apparent to the naked eye: they are slightly out of step with each other. Let us assume that reflection at a mirror or a beam splitter changes nothing about the ray but its direction. (This is a slight oversimplification, but it will do.) Transmission through the metal coating of a beam splitter, by contrast, delays the crests of the light wave by a quarter of a wavelength, a wavelength being the distance from one crest to the next. With this difference in mind, notice that the two rays labelled rt and tr both suffer exactly the same delay, a quarter of a wavelength from the original, because each one was transmitted only once. These two beams (rt and tr) fall neatly on top of each other and interfere constructively. Therefore the symmetric meter S measures the full beam of light.

The other two rays, rr and tt, are different. The twice-reflected ray rr is completely unaffected by its journey. The twice-transmitted beam tt, on the other hand, is delayed by two quarter wavelengths, or half a wavelength. This means that the crests and troughs of the beam called tt have reversed their positions! Consequently rr interferes destructively with tt, cancels it out, and causes the meter U to remain in darkness.

The cleverness of the coupling of beam splitters into an interferometer might suggest that interference is a rare phenomenon, but actually, quite the opposite is true. Interference is the hallmark of waves, and occurs wherever two wave-trains meet. Water waves, sound, light, microwaves, X-rays, seismic waves and wiggles on a stretched piano wire all display interference. Indeed, nature delights in it. Its effects are seen most impressively in the colours of peacock feathers and of butterflies, and most delightfully in soap bubbles, whose flimsy walls act as miniature beam splitters. A small part of a ray of sunlight (only about 4 per cent) is reflected from the front surface of a soap film, while the rest travels through the water to the rear surface, where another tiny portion is reflected. (Most of the light continues straight through to the other side.) Owing to the shift suffered during transmission through water, the two reflected beams, when they meet again as they travel together toward your eye, may interfere. In particular, depending on the thickness of the film, the second ray may be delayed with respect to the first one. For some colours the interference is constructive, for others it is destructive, so different colours are enhanced and suppressed by different portions of the soap bubble.

Returning to our Mach-Zehnder interferometer, let's consider one last refinement. Suppose the auxiliary mirror in the lower path (labelled r) is equipped with a knob that allows its position to be shifted by minute amounts. This changes the length of path r of the once-reflected beam, causing an additional shift in the light in that branch. If the change happens to be through half a wavelength – the distance from a crest to the nearest trough – crests and troughs of that ray will be interchanged at the merging point. The result will be an exchange of the responses of the detectors U and S.

Now you are ready for a neat experiment. While slowly altering the path length of r, watch the intensities of light measured by U and S. First, as before, the full intensity arrives at S. Then, as you adjust the mirror, U begins to register a signal, while, conversely, the brightness of S diminishes, until at some point each detector receives half the light, and finally S itself turns dark. Interestingly, further adjustment of the mirror brings the reading of S back to its original full value – a cycle that is repeated smoothly and regularly, back and forth, and, plotted on graph paper, traces out two complementary sinuous curves. Where one rises to a maximum, the other sinks to zero, and their sum always adds up to the intensity of the incoming beam. In each detector, constructive and destructive interference alternate in regular succession. Classical optics, and the wave theory of light, effortlessly account for the entire sequence of events. Baseballs and buckshot, on the other hand, do not interfere, and where they end up most certainly does not depend on how far they have to travel before they get there.

If the intensity of the incoming light is turned way down, and light meters are replaced by photon counters, the whole sequence will take longer, because the clicks of the counters need time to add up to significant numbers. The final results, however, are the same as before: *the clicks prove that light is made of particles; the interference proves that it consists of waves*. Thus the interferometer equipped with photon counters demonstrates the quantum mechanical wave-particle duality of light.

The action-at-a-distance of quantum mechanics, which Einstein called 'spooky', can also be illustrated by the interferometer. Imagine a single photon in the transmitted path t of the first beam splitter. If the reflected path r is open, the photon, as we have seen, will never reach detector U. But if path r is closed, the photon can trigger a click in detector U. Thus *closing* one of the paths can result – paradoxical as it may seem – in an *increase* in counts of the U detector. More disturbingly, one may ask: How is the photon in path t to 'know' whether r is open or closed? On the scale of sizes of atomic particles, the two paths are very far apart – a few inches in a laboratory interferometer, a dozen miles in some recent experiments in Switzerland. Somehow the photon seems to be in two places at once – in path t in order to pass through the apparatus,

in path r to check whether or not that channel is ope
light is described in good nineteenth-century tradition as
the answer is by now familiar: a wave *can* be in two places a
and if you block one of them, destructive interference canno...
place; but a single photon is indivisible – no detector has ever
measured the energy equivalent of half a photon!

If there is a lingering doubt in your mind that photons are
particles, you are in good company. They are certainly not ordinary
particles that can be manipulated and stored like grains of sand or
pellets of buckshot; but all the weird quantum phenomena we
have seen illustrated by photons apply equally to particles with
impeccable material credentials. When the incoming beam of light
is replaced by a beam of solid, tangible particles like electrons,
neutrons, or even entire atoms or large molecules, and the mirrors,
beam splitters and detectors are suitably modified, the results are
exactly the same as for photons. This, then, is the mystery
(Feynman's 'only mystery') of quantum mechanics: *All particles in
nature are sent randomly into one of two directions by quantum mechanical beam splitters, just as photons are. All particles in nature display
the phenomenon of interference, just as waves do. All particles in nature
experience spooky action-at-a-distance.*

This scenario is truly mind-boggling. Picture this: every morning
a girl enters a walled garden along the same path. She soon comes
to an intersection where another path crosses hers. Being of an
equable temperament, she chooses randomly from day to day
whether to stay on her original course (labelled t) or to switch over
to the other path (labelled r). A little further on, the same two
paths cross again on their way to two exit gates (labelled S and U).
The two intersections look exactly alike and represent two identical
beam splitters. Amazingly, every day the girl emerges from the
garden through gate S – never through U. However, if one of the
two paths – it doesn't matter which one – happens to be closed for
repairs at some spot between the intersections, she comes out of
the two gates in random alteration – as might be expected.
Somehow, the availability of the path not taken has a decisive
effect on the girl's passage.

We have seen how waves can achieve these feats by virtue of interference; but how a material particle manages to behave in such a bizarre manner is truly mysterious.

Waves are absolutely necessary for explaining interference, yet particles of matter are not waves. How do physicists bring the understanding gained from optics and other wave phenomena to bear on the description of particles? The invention that does the trick is a mathematical expedient called the 'wave function', the conceptual heart of atomic physics and chemistry. The graphical representations of the wave functions of modern science resemble ocean waves for particles propagating freely through space, and exquisitely complicated cloud formations for electrons confined in atoms. The wave function, which Erwin Schrödinger invented expressly to make the interference of material particles like electrons comprehensible, is not real. It is associated with a particle in the same way that your National Insurance number is associated with you. It helps to keep track of things, but has no independent physical reality the way your hand does, or your wallet. The wave function is an abstraction, but, like your NI number, it too has powerful consequences when used with a purpose.

The rule that ties an object, such as an electron, to its wave function reflects the essential nature of the quantum formalism. On the one hand, the wave function is perfectly predictable, but on the other hand, there is something irreducibly random about the behaviour of an individual electron, as a humble beam splitter demonstrates. How can both of these seemingly contradictory statements be true? How can predictable mathematics describe the essential unpredictability of nature?

Quantum mechanics reconciles the irreconcilable by postulating that the wave function doesn't describe the electron itself, but merely the probability of finding it. (Three-quarters of a century of ingenuity, effort, controversy and doubt lie hidden behind that innocent little sentence!) An electron is real; a probability is not. The quantum wave function, together with its interpretation as a probability, thus accounts for both the interference of particles and their random behaviour. The relationship between the predictability of the wave function and the unpredictability of the electron is analogous to the relationship between the two seemingly contradictory statements: 'The probability of throwing heads, a perfectly predictable quantity, is precisely 50 per cent,' and 'Whether this penny will come up heads or tails is utterly unpredictable.' The contraption called 'wave function', which achieves this compromise, seems both contrived and bizarre, but it works.

The miracle is not why nature behaves in such a way – it simply does, and we should be grateful that we have been allowed to figure out this much. The miracle is *how* we were able to gain so much insight into the inner workings of the atom, at a scale of distances so far below our own. As we saw earlier, Einstein's optimistic remark that God is subtle but not malicious found some corroboration in the linear structure of chromosomes, which allows the genome to be mapped. In the case of quantum mechanics, it turns out that while electrons and nuclei are hidden deep within matter, nature has arranged for one atomic, elementary particle that we can see with our own eyes and manipulate with ease: the photon. Its dual nature as a particle and a wave is put into bold relief by the beam splitter and interferometer. The enigmatic photon is nature's messenger from Lilliput: it allowed us to figure out how the world operates in the atomic realm.

Modern quantum theory was born when the French physicist Louis de Broglie suggested, following nothing more solid than his intuition, that if light (a wave) could behave as though it were composed of particles (photons), then the electron, which is manifestly a particle, might likewise have wave properties. This happy guess turned out to be right, not only for electrons, but for elementary and composite particles, all the way up to complex molecules. Even lead pellets and mice can, in principle, display interference, although we may not be able to catch them at it for a long time to come.

Randomness and interference are the hallmarks of quantum behaviour. The wave function incorporates and quantifies them both. But what *is* the wave function? Its interpretation as a probability provides a clue. Probabilities, as we saw, are shorthand expressions of partial knowledge – of information. Similarly, the wave function of an electron in an atom contains information about the likelihood of finding the electron in various places. It is a map of potentialities – a catalogue of possibilities. It is information, pure and simple; and its shape determines the form of the atom.

The hauntingly strange architecture of the world at its most basic level features graininess where smoothness was expected, randomness where predictability should reign, spooky action-at-a-distance, and the interference of matter described as wave-particle duality; but the weirdest quantum effect of all has no analogue whatever in the everyday world. It is called 'superposition' and plays havoc with the concepts of truth, falsehood and information.

The wonder of quantum superposition

Schrödinger's cat is the most famous feline in physics. Since its birth in 1935 it has provoked endless and occasionally violent debates among physicists. (I recall a lecture in which Stephen Hawking, through his touch-activated voice synthesizer, grumbled: 'When I hear mention of Schrödinger's cat, I want to reach for my gun.') In the popular imagination it has come to be identified with the most bizarre, counterintuitive aspects of modern physics, and inspired books, essays, cartoons and poems. Although the cat has surely exceeded its creator's fondest hopes as a stimulus to thought, it remains an abject failure as an aid to public understanding of science.

The set-up is simple enough. Imagine a normal house-cat in a windowless, soundproof cage. Place a radioactive nucleus, in its original, undecayed state, into one corner of the box, rig a Geiger counter next to the nucleus so that it can detect the decay when it occurs, and connect the output signal from the Geiger counter to an electrical gizmo that opens a vial of poisonous gas capable of killing the cat instantaneously and painlessly. Shut the cage door and wait.

Without opening the box, you cannot tell what is going on inside; but classically applied logic tells you that, after a set period of time, the nucleus either will have decayed or will remain undecayed – that something will or won't have happened inside. Simple enough, you might think; but this statement is just plain wrong. For, as long as you don't peek, the nucleus actually enters a curious quantum-mechanical state called a 'superposition', in which its status is described by a wave function. Under these conditions – and not peeking is key here – the nucleus is not one thing or the other; it is both: rather than being described as 'decayed *or* undecayed', it must be 'decayed *and* undecayed'. Despite the apparent counterintuitive nature of this new statement, superpositions are common in nature, and indeed we have already encountered one. A particle that hits a beam splitter, for example, is at once 'transmitted *and* reflected'. If it were not – if it behaved like a miniature marble that is 'transmitted *or* reflected' – the interference that is so boldly exhibited by an interferometer could not occur.

However, if the nucleus in the box is at once 'decayed *and* undecayed', the cat must be simultaneously 'dead *and* alive'. As soon as you open the box, you will find the nucleus either whole or disintegrated, and correspondingly the gas either in its vial or dispersed, and the cat either dead or alive; but until then the contents of the cage are in an intermediate state of a kind we do not recognize in normal daily life. Schrödinger knew very well that this conclusion defied common sense, but offered no satisfactory way out of the dilemma. Over the years, countless ingenious resolutions have been proposed, but none have been sufficiently convincing to end the debate.

The trouble with the story of Schrödinger's cat is that it's all in the mind; before anyone has looked, its status (whether it is labelled dead or alive, or both, or halfway in between) matters to no one, least of all the pussy. What is needed to bring the notion of superposition more forcefully into public consciousness is a persuasive, in-your-face kind of argument, one with real consequences that can't be dismissed as idle philosophizing. One such story is called 'A Game of Beads', and I am grateful to Lev Vaidman of Tel Aviv University for relating it to me one evening over a glass of wine in the Schrödinger Institute of Physics at the University of Vienna in Boltzmann Lane.

The game pits two teams, consisting of two classical physicists playing against two quantum physicists, to see which team can best fool an interrogator. The interrogator has asked the teams to perform an apparently simple task: string an *even* number of green and red beads onto a round necklace in such a way that adjacent beads always have different colours while the two end beads, next to the clasps, have the same colour. In other words, if the first bead is green, the next should be red, the one after should be green, and so on till the last bead, which should be the same as the first. Imagine using six beads and you will quickly realize that, with an even number, this scheme is impossible to carry out.

Fortunately, though, the rules of the game don't require that the players actually present the interrogator with such a necklace; they must merely answer questions about adjacent beads, even bluffing if it suits their purpose. The interrogator asks each team to create two identical necklaces, and orders the four players to repair to four separate soundproof booths with one necklace apiece. Each player is asked a single question. The interrogator might ask a player: 'What colour is the fourth bead?' and ask the second player of the same team about the colour of an adjacent bead, either the third or the fifth in this case. If the answers turn out different, the team passes the test; if they match, the team fails. (For the beads at the clasps, the answers must, of course, match to pass.) The two questions are then repeated for the second team. If both teams pass – or both fail – the round is a tie and new necklaces must be produced for another round.

The two teams contribute equal amounts of money, say £1000 per round, to a betting pool. The winner is the first team to pass a round while the other one fails. The money piles up quickly!

Under ordinary circumstances, the best each team can do is make two copies of an actual necklace with a single fault in it – a pair of same-coloured neighbours or different-coloured end beads – and answer the questions honestly. Without having the actual necklace, or at least an accurately remembered design for one, a player would not be able to guess what their partner might say, and hence what answer to give. Presuming that a team does have a pair of necklaces in their possession, their best chances for success with six beads would be no more than 5 in 6, or about 83.3 per cent.

Now for the clever bit. Imagine that the quantum team tries to get round the need for making or imagining actual necklaces, and decides to use a quantum-mechanical device in their place – based on a pair of electrons. An electron is a tiny magnet, described by an imaginary arrow called its spin, which has a curious quantum-mechanical property. If you pick any direction you wish, and decide to measure the direction of the spin, you will find that it points either along the direction you picked (spin up), or the other way (spin down). Somehow, the spin arrow is simply not allowed to point at 30 degrees from the axis along which you aligned your measuring device – or in any other direction except parallel or antiparallel. Furthermore, if you produce two electrons together in an atomic reaction, you can arrange to couple their spins so they are opposite each other, like two bar magnets joined North to South and South to North. In that case, nothing is known about the direction of spin of either particle. If one of the spins is measured, it will have a definite value – say 'up' or 'down' – and then the other particle, even if it has been removed to the other side of the galaxy, will with certainty have the opposite spin when measured along the same direction. The two particles are said to be *entangled* (a term coined by Schrödinger) and are described by the superposition '#1 is up while #2 is down' *and* '#1 is down while #2 is up'. Which of these two possibilities will actually show up in a real measurement is as unpredictable as the pulse of a cat that happens to be dead and alive.

Back to the contest. Before each round, the quantum team produces two entangled electrons and carefully separates them so that they remain entangled. Each player takes one electron into the interrogation booth. They agree to measure the spins of their particles along certain pre-selected, numbered, equally spaced directions that correspond to the numbers of the beads. In the case of a six-bead necklace – where the directions would differ by 30-degree steps from 30 degrees to 180 degrees, halfway around a semicircle – the first player would ascertain the colour of the fourth bead by measuring spin in the 120-degree direction. He would answer red or green, depending on whether the spin was up or down. The second player, upon being asked, 'What is the colour of bead 5?' measures the spin of the entangled particle in the 150-degree direction and answers accordingly. Since the delicate relationships between electrons are destroyed by the act of measurement, a new pair of entangled electrons is required for each round of four questions.

Notice that there are no beads, no colours, no necklaces – not even designs for a necklace. Nor are the particle spins pointing in definite directions until the measurements are performed. The goal is not to make an impossible necklace, nor even a quantum analogue of a necklace, but to fool the interrogator.

The probability of finding opposite colours for adjacent beads, or similar colours for the two end beads, can be calculated by the rules of quantum mechanics and comes out to be about 93.3 per cent, substantially higher than the value of the probability derived from actual beads. For a game lasting several rounds, that advantage accumulates with each round. As the number of beads increases, the quantum-mechanical advantage over the classical strategy rises dramatically. In fact, for long necklaces the probability that the quantum team will win the game rapidly approaches certainty. Since no one could ascribe this phenomenal success to luck, an interrogator ignorant of quantum mechanics would be compelled to believe that there really are impossible necklaces behind the quantum team's closed doors – or else that the quantum contestants are communicating by telepathy.

The cunning of the game lies in the selection of directions along which spins are measured. For a large number of beads, two adjacent directions *almost* coincide because the division of a semicircle into a large number of equally spaced directions produces ever narrower slices; but in the entangled state in which the particles have been prepared, the spins measured along any line point in opposite directions, so along two adjacent directions they will *almost* always differ. The only exception occurs at the ends. The two directions corresponding to the first and last bead are *almost* 180 degrees from each other. For the special entanglement considered here, spins measured along *opposite* directions must always be both 'up', or both 'down' – hence calling for beads of the same colour. In this way the quantum system neatly mocks up a necklace with just the right combination of properties.

The outcome of this game is not in doubt. Although the predictions of quantum mechanics, such as the formula for the probability, were debatable in 1926, enough of them have been corroborated experimentally in the meantime to make quantum mechanics as robust as car mechanics! To be sure, it is not easy to prepare entangled particles, or to measure their spins in precisely defined directions in different rooms; but there is no disagreement among physicists about what will happen when the technology finally catches up with the theory.

It is important to realize that the entangled particles are not some kind of signalling device: there is no communication between the players. Instead, there is a correlation between their measurements. For the classical team, a correlation is embodied in their real, identical necklaces. Quantum correlations are very different from classical correlations, and are sometimes called *stronger*; but that word does not really do justice to their counterintuitive nature.

The Game of Beads enjoys a twofold advantage over the story of Schrödinger's Cat. First, it is directly tied to specific, measurable outcomes which will eventually be realized in a laboratory experiment. More important, though, is the fact that there is a real consequence. The game matters: money changes hands. Both stories strive to connect an atomic system ruled by quantum theory with the ordinary, everyday world. But whereas the radioactive nucleus is connected to the cat via a Rube Goldberg device, the link between the entangled particles and the contestants who use them is logical. It is at once less tangible and more sturdy, like the way in which the throw of a die can have serious, even fateful, consequences for a bettor.

And now to the moral of the story. The lesson of the Game of Beads is that the unexpected correlations between the measurements of spins are only possible if the individual particles exist in states of superposition: not up *or* down, but up *and* down. Transposed into our world, that means that each bead of the fictitious quantum necklace is both red *and* green in some profound sense.

Imagining a bead that is red and green at once is as difficult as trying to imagine the fourth dimension. I envision a quantum bead as a beautiful but indistinct, shimmering, translucent and psychedelic sort of object that freezes into one specific colour the moment I touch it; but I don't put much faith in that image. I trust the power of science, and the ability of my mind to grasp abstract concepts. My brain has shown me that the foundations of the world are really completely quantum-mechanical. If it were not for the crudeness of our senses, we would know from experience that beads can be red and green, that spins can point up and down at the same time, that electrical currents in a wire loop can flow simultaneously in opposite directions and that, as the Canadian humorist Stephen Leacock put it, a rider can jump on his horse and ride off in all directions. The question is no longer: how can that be? The burning question for physics has become: how come we don't notice superposition in everyday life? or: how are the superpositions of the world at its fundamental level disguised to yield the stark outlines of the world our senses perceive? Perhaps Wheeler's really big question WHY THE QUANTUM? should be rephrased as WHY THE CLASSICAL?

The quandary is reminiscent of another one that is as old as the atomic doctrine: how is the graininess of the world at its fundamental level disguised to yield the smooth outlines of the world our senses perceive? The answer to that question is easy: it hinges on the smallness of atoms. Today the scanning electron microscope and its descendants have enabled us to plunge down into the atomic realm where graininess rules. Atoms have become commonplace. In an analogous fashion we will, in the coming century, magnify the weird aspects of reality to such an extent that our intuition is re-schooled. Then we will stop eyeing superposition as though it were a logical impossibility, and embrace it in the way we have accepted atoms. The sooner we adopt a world picture that fully incorporates the real phenomenon of superposition, the more quickly we will be able to reap the technological pay-offs it is capable of returning.

Nowhere is the difference between either/or and both/and more clearly apparent than in the context of information. To be sure, the binary choice between zero and one is a question of logic, not of physics; but since, as Rolf Landauer insisted, information is physical, the phenomenon of superposition must change the rules of information-processing in a radical way.

21 The Qubit

Information in the quantum age

Inasmuch as *or* describes the state of a bit – zero *or* one, true *or* false, yes *or* no, dash *or* dot, red *or* green – it is obvious that if quantum mechanics is brought to bear on information theory, as it is on every other branch of physical science, or if information is stored in a quantum system, as it must be if the dimensions of computer memories continue to shrink, something has to give. With acceptance of quantum theory comes the realization that the blending of mutually contradictory attributes is the normal state of affairs among atomic systems. For, as we have seen, at the most fundamental level things are both here *and* there, up *and* down, particle *and* wave, red *and* green. The conjunction 'or' kicks in only when interventions occur in the form of measurements or observations – when the choices imposed on quantum-mechanical systems force them to look and behave more like the familiar objects of our sense experiences. Thus, when the crisp, exclusive 'or' gives way to the dithering, inclusive 'and', information theory necessarily goes with it; and yet, it was almost half a century after Shannon's pioneering paper before the quantum version of the bit was invented.

In May 1992 William Wootters, a physics professor at Williams College in Massachusetts, who had studied under John Wheeler and helped to lay the groundwork for the theory of quantum information, was ending a visit to his colleague Benjamin Schumacher at little Kenyon College in Ohio. On the way to the airport the two joked about nomenclature: maybe communication about their slippery subject would be aided by a new term for a quantum unit of information. They fooled around with phrases such as 'quantum bit' and 'q-bit', finally settling on 'qubit', which they pronounced with the accent on the first syllable, like the archaic measure of length called the 'cubit', but bearing no relationship to it. After their laughter ended, the idea lingered, and in the autumn Schumacher formally proposed the new word to a grateful scientific community, who adopted it at once. Today, a decade later, it is impossible to imagine the burgeoning literature on quantum information without the qubit.

As the offspring of the union of a bit with a coin toss, of certainty with randomness, the qubit has inherited certain properties from its parents; yet it differs substantially from both. Like both, it is a concept, not a thing. Where a bit can have the value zero *or* one, a qubit is defined as a quantum superposition of zero *and* one. When its value is actually measured, it will be zero or one at random, just like a coin toss; but before measurement, the qubit can exist in a balanced state of 50 per cent *one* and 50 per cent *zero*, or in any other weighting, such as 80 per cent *one* and 20 per cent *zero*, like a crooked gambler's biased coin spinning in the air.

As a consequence of this, measurement, whether by human or mechanical intervention, must in some sense be considered a creative act. Before it takes place, there exists potential information. Afterwards, a definite, unique bit of information has been created that did not exist before. It's a process comparable to the act of writing – fixing the multi-valued, indistinct, ambiguous and indefinite words flitting through my brain onto my monitor in definite form.

The manipulation of qubits reminds me of an old voting method still in ceremonial use at the official meetings of the founding chapter of Phi Beta Kappa, the academic honour society, at my university. When the selection of new members for the chapter is in order, an ancient voting machine is fetched from storage and placed on the table in front of the presiding officer. This antique piece of furniture is a dark mahogany box about the size of a six-pack of beer, highly polished and adorned with elaborate mouldings. Through a hole just big enough to accept a human hand you can see a felt-lined cavity inside the box, with two open compartments on its floor. Numerous black marble-size balls roll around in one of them, white balls in the other. To vote, a member reaches one hand into the hole, picks a white ball from the right-hand side to signify yea, or a black ball from the left-hand side for nay, and pushes it through a small opening in the back of the cavity, whence it drops into a drawer at the base. Since this action takes place inside the box, and the voter's wrist blocks the line of sight, it is hidden from view. When the box has made its way around the table, the drawer is opened, and the black and white balls are counted. A potential candidate for membership rejected by this method is said to be 'blackballed'.

Used in this context a qubit would be a token that is simultaneously black and white – a changeling, like the red and green pellets in the Game of Beads. When it plops into the drawer it acquires a definite colour, but until then it remains in an indeterminate state. This magical ball would be useful for a voter who feels torn between yes and no, and wishes to leave the outcome to chance. (More refined qubit balls could represent different weightings of opinion between the two extremes of yes and no. They would be picked by a voter who felt that the rationally defined, Bayesian assessment of the worthiness of a candidate were, say, 80 per cent.) So long as the members of Phi Beta Kappa are discreet, exactly what happens inside the voting box remains mysterious. Such a device, which has clearly defined inputs and outputs, but no explanation of what goes on inside, is a common metaphor of physics and is called a 'black box'. In this sense the entire machinery of quantum mechanics is something of a black box, because on its inputs and outputs there is universal agreement, whereas its mechanism remains debatable.

For voting, a single qubit ball is of marginal utility. If the choice between black and white is to be left to chance, a voter could just as well fill the voting machine with an equal mixture of black and white balls, shake them out of their compartments to mix them up, and then reach in and select one blindly; but while the outcome of this procedure is identical to what happens if a single qubit ball is dropped into the drawer, the explanations of the two operations differ. In the conventional, so-called classical case, the balls have well-defined colours all along – we just don't happen to know what they are until the drawer is opened. In the quantum-mechanical case, on the other hand, the qubit ball has no definite colour to start with. There is no way, even in principle, to determine whether it is black or white before it is dropped into the drawer. It is this essential indeterminacy that distinguishes the qubit from the bit.

Things get more interesting when a second qubit ball is added to the first. By means of certain simple quantum-mechanical operations, two qubits can be entangled. (A pair of entangled qubits is like an old married couple whose habits are so well established that, when one is observed, the behaviour of the other one can be predicted.) With a touch of legerdemain a voter could prepare two qubit balls in such a way that while their colours would remain unpredictable, they would nevertheless definitely be the same. As soon as the first qubit ball is found to be black, for example, the second, entangled one will of necessity be black also. In this way one vote, indeterminate as it started out, doubles in value. Alternatively, if the two qubit balls are in an entangled state of opposite colours, as in the Game of Beads, a vote with one ball will always neutralize the other one. A quantum computer, it will turn out, could be viewed as nothing but an immense elaboration of a Phi Beta Kappa voting machine with qubit balls.

That image is too crude, however. A qubit, after all, is a concept, not a thing, so the qubit ball is a fiction, and an unconvincing one at that. Since qubits are fundamental not only for understanding information in the quantum age, for quantum computing and ultimately for describing physical reality, it is worthwhile to follow a real qubit from creation to measurement.

Fortunately it turns out that a beam splitter, that quintessential quantum device, happens to be the ideal machine for making qubits.

Consider a beam splitter mounted horizontally, its face parallel to the ground. If a beam of laser light impinges on it obliquely from above, some of the beam is transmitted through the glass and its metallic film to the space below, while some of it is reflected back up to the space above. Now suppose we turn the intensity of light way down to the level of photons, and agree that each photon is to carry one bit of information. The coding convention will be this: a photon that ends up *above* the beam splitter means 1, a photon that ends up *below* means 0. In this simple way we have constructed a qubit generator.

Imagine launching a single photon carrying the message '1' onto the beam splitter. The result is that there will be two signals, one above and one below. However, neither a second photon, nor a second bit of information, has been created. According to what we learned in the last two chapters, the information in the machine now resides in a quantum-mechanical superposition of 1 and 0. If two detectors, above and below the beam splitter respectively, were to intercept the signals, one of them would fire, but it would be impossible to predict which one. This is the crucial clue that a qubit '1 *and* 0' has, in fact, been created. It is carried by two feeble beams of light, which together carry just enough energy for a single photon.

As with all quantum devices, a qubit is a delicate flower. If you so much as look at it, you destroy it. Unlike a robust bit, which continues to have the same value no matter how often it is examined, a qubit loses its identity when its value – zero or one – is measured. Nevertheless, there are ways to manipulate a qubit. In the first place, the beam splitter need not be 50/50. Instead of reflecting half of the incoming beam and transmitting the other half, it can be designed to reflect any portion of the total between 0 and 100 per cent. For our purposes, however, we will usually stick to symmetric beam splitters and 50/50 qubits.

A different adjustment relies on the wave nature of light. By modifying the path lengths of the two emerging beams, the relative positions of their crests and troughs can be varied. This means, in turn, that when the two beams are brought together again, or blended with other beams belonging to other qubits, interferences ranging from destructive (crest meets trough) to constructive (crest meets crest) can be achieved. The variability in relative weights and relative path lengths together make the beam splitter a versatile tool for generating an infinity of different qubits.

If a qubit is made from a photon above the beam splitter, which, as we agreed, carries the information '1', what happens to that definite bit of information? It is not lost or destroyed in the course of the creation of the qubit, but hidden in a subtle way. It is spread out over the two light beams (one reflected, the other transmitted) and can only be recovered if these are both considered together. One possibility for accomplishing this is to shine them onto the upper and lower surfaces, respectively, of a second, identical beam splitter, as we did in constructing an interferometer. There it emerged that destructive interference occurs above the second beam splitter (rr and tt are out of synch), and constructive interference below (rt and tr are in synch). The outgoing photon will therefore with certainty carry the message '0'. (Conversely, if the initial incoming bit is '0', the locations of constructive and destructive interference will trade places, and the outgoing message will be '1'.) Thus a simple reversal of the symbols '1' and '0' will occur, and this entails no loss of information. Communication with a compulsive liar, who systematically interchanges Yes and No, is as reliable as communication with a habitual truth-teller.

A device that switches 1s and 0s, the way the interferometer does, is called a '*not* gate'. (Logical circuit elements are called gates because they control the flow of information the way adjustable gates control the flow of water through irrigation ditches.) Since there are only two possible values to a bit, rejecting one means perforce choosing the other, so in this context switching is equivalent to negating. *Not* gates are common components of ordinary computers. If the two possible values of a bit are carried by two wires, a *not* gate is easy to design: just cross the wires, connecting an input 1 to an output 0, and vice versa. A child could do it – in the end an interferometer, a sophisticated instrument consisting of two beam splitters and two mirrors all carefully aligned, accomplishes the same goal as a couple of crossed wires. The optical device appears to be a waste of ingenuity.

The point, as with qubit voting, is that crossed wires carry bits from beginning to end, but an interferometer translates them into qubits before recovering them. It is the intermediate state that is of interest. Schematically, an ordinary circuit consists of three elements: initial bit, *not* gate, final bit. A quantum *not* gate, in contrast, has five elements: initial bit, beam splitter, qubit, beam splitter, final bit. Quantum-mechanical manipulations can be performed on the qubit, with results that are impossible to reproduce with currents and wires.

A single beam splitter, considered by itself, is, in fact, a '*square root of not*' gate. The terminology arises from the fact that, applied twice, it results in a *not* gate, just as the square root of 2, multiplied by itself, equals 2. Remarkably, a *square root of not* gate cannot be constructed from ordinary wires. (Try it!) There is no classical circuit which, repeated twice, produces a simple *not* gate – just as there exists no logical operation which, twice repeated, yields a negation. The *square root of not* is a uniquely quantum-mechanical concept.

The *square root of not* is the logical counterpart of the mathematical 'square root of minus 1', which is called imaginary and universally abbreviated 'i'. Of course 'i' is a perfectly straightforward mathematical concept, and no more imaginary than the concept of love, but it is definitely not a real number. That there should be a relationship between a simple quantum gadget, such as a beam splitter, and the imaginary numbers of algebra, hints at deep connections between physics and mathematics.

The analogy with imaginary numbers suggests a helpful geometrical representation of a qubit. Negation, in this context, is the operation of passing from the bit value 1 to the bit value 0. If these two extremes are represented on the surface of a globe by the North and South Poles respectively, what is the meaning of a trip from one to the other over the surface of the globe, through London, say, and across the equator? The intermediate steps may be interpreted in the quantum-mechanical sense of superpositions of both bit values. The operation *square root of not* then means travelling south a quarter of the way around from the North Pole

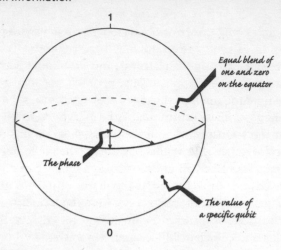

to the equator. Repeating this step gets you halfway around to where the opposite resides.

If latitude represents the relative proportion of zero to one in the superposition, what does longitude refer to? The second quantum-mechanical parameter of a qubit that is not apparent in an ordinary bit is the phase – the degree to which its two waves are out of step. At the Greenwich meridian, with zero longitude, the two are perfectly in synch, but on the other side of the globe, in the middle of the Pacific, they are 180 degrees, or half a wavelength, out of step. (Since the two waves remain separated as long as they are in a qubit state, they do not interfere. That happens only later, when they are brought together again.) At the North and South Poles longitude is irrelevant because the qubit is not a superposition in those two places. Everywhere else the two numbers, relative proportion and relative phase, latitude and longitude, completely specify the qubit.

In much of the popular literature about quantum computing, a qubit is represented by a little ball, with 1 at the north pole and 0 at the south pole. It's a nice, reassuring image, connoting unity, compactness and simplicity. It opens the possibility of an infinity of logical choices intermediate between true and false, between a proposition and its negation. We have quantum mechanics to thank for suggesting this radically unconventional way of parsing reality.

If qubits can carry hidden information and allow additional manipulations that are impossible to perform on bits, they also suffer from a fundamental disadvantage vis-à-vis bits. In 1981, long before the qubit was invented, Wojtek Żurek, who improved on Boltzmann's entropy formula with the help of algorithmic complexity, together with William Wootters, the godfather of the term 'qubit', established an unexpected consequence of quantum theory called the 'no cloning theorem'. It stipulates that an unknown quantum state – such as, for example, a qubit – cannot be copied without destroying the original. A bit, in contrast, can be measured and copied at will without degradation.

Of course, if the two parameters – the longitude and latitude – of the qubit are known, many copies of it can be constructed. However, without this prior knowledge the qubit cannot be reproduced. If it could, complementary measurements could be performed on two identical copies of it, and Heisenberg's uncertainty principle could be circumvented. Since this is impossible, cloning must also be forbidden. Both the no cloning theorem and the uncertainty principle are fig leaves with which nature protects its quantum secrets from prying eyes.

As carriers of information, the bit and the qubit differ in several respects. A bit represents a single binary digit – the answer to a yes-or-no question. A qubit, on the other hand, is not fully specified until its longitude and latitude are fully described, out to an infinity of decimal places. Thus a single qubit could theoretically be used to carry an unlimited amount of classical information, just like Champ's magical Constant and a ring with a precisely known radius. Two practical problems interfere with the enticing prospect of using qubits as compact computer memories. For one thing, a qubit, like a carefully turned ring, is limited in accuracy by error – or 'noise' – in the process of preparation. Furthermore, since the information carried by a qubit cannot be read off directly, but only inferred from the probability of finding the result 1 or 0, and since that probability cannot be well determined without a huge store of identical copies of the qubit for use in repeated measurements, qubits offer no advantage over classical bits for data storage.

Another difference between bits and qubits lies in their relationships with each other. Two bits can be related via the meaning of the message they convey in juxtaposition, but physically they are always represented by different material entities. Two qubits, on the other hand, even those that are physically distinct, can be connected via the infinitely elastic, invisible strings of entanglement. Thus qubits offer richer, more complex opportunities for processing information.

How to stuff bits into qubits and squeeze them out again without losing them to the inevitable noise, how to classify and measure the degree of entanglement among multiple qubits, how to move classical and quantum information through wires and through the impalpable connections among qubits – these are some of the hard questions that occupy modern quantum information theorists. The answers, it is hoped, will not only illuminate the nature of information, but may eventually be put to practical use.

22 Quantum Computing

Putting qubits to work

Right from the beginning, quantum mechanics has suffered from a disquieting case of split personality. Despite its central role in shaping modern science, which extends far beyond the theoretical to the practical, its youthful and robust public face has another side to it. For, in private, quantum mechanics is something of a delicate, fey neurasthenic, one anxiously watched by a small army of theoreticians, mathematicians and philosophers who worry about its state of health if not its very survival. Quantum computing, its latest offspring, has inherited the same disorder. According to the popular press, quantum computers promise to deliver calculations of unimaginable complexity – and much more besides. Inside the profession, however, there are those who insist that quantum computing is not just hyperactive conventional computing, but different in kind; they burden it with heavy metaphysical responsibilities and worry about whether it can carry them out. The late Rolf Landauer, for example, who discovered the cost of forgetting, implied that our world view depends on it: 'Not only does physics determine what computers can do, but what computers can do, in turn, will define the ultimate nature of physical law.' Poor quantum computers. Before they are even out of their nappies they are expected to carry the entire edifice of physics on their puny shoulders.

Originally, quantum mechanics was believed to be a hurdle in the way of the exponential expansion of information technology – a potential limit to growth. As computations grow more massive and computers shrink in size, it is not difficult to foresee a time – the decade of the 2020s has been suggested – when memory cells reach the one-atom-per-bit level. At that scale, classical physics makes way for quantum physics. Would quantum mechanics, with its inescapable elements of uncertainty and randomness, in turn, smudge the carefully drawn blueprints of computer logic, and put a stop to the shrinking trend of memory cells? The question was at the forefront of fundamental computer science in 1985 when David Deutsch, a theoretical physicist at Oxford University, turned it on its head. Far from limiting computation, he pointed out, quantum mechanics can enhance it in a spectacular way.

Conventional computers perform their operations by the straightforward, step-by-step method we learned for long multi-plication in junior school, a technique called *serial computing*. When several computers share the work of a calculation – for example by multiplying units, tens, hundreds and thousands sep-arately and adding all the results in the end – the time it takes is divided by the number of participating machines, and the method is called *parallel computing*. Large computations, such as those encountered in numerical relativity, rely on this technique to keep the time requirements and costs within manageable bounds. Quantum computers, however, have the potential to do much better, thanks to a revolutionary approach Deutsch called *massive parallelism*.

In an ordinary computer, each bit of information, each 0 or 1, is stored in a separate memory cell; so with three available cells, the maximum that could be stored is a single three-bit string. In a quantum computer, on the other hand, each cell contains a qubit, which consists of a superposition or blend of 0 *and* 1. Three cells are therefore capable of storing not just one three-bit string, but all eight possible three-bit strings at once. If we were to examine the contents of those three cells, we would find one of the eight strings – say 101 – at random, and the other seven would be irretrievably lost. The result would be useless; but fortunately such a move is not always necessary.

Being very careful *not* to make a measurement, we can manipulate the three qubits quantum-mechanically. They evolve smoothly, following the commands of Schrödinger's wave equation, into some new configuration. In this way we can perform a calculation with eight strings as input in the time it takes an ordinary computer to do the same thing with a single three-bit input string – and the bigger the number of qubits involved, the bigger the advantage of massive parallelism over its clunky conventional cousins.

So far so good, you might think – but there's a problem. Say a calculation has been performed on one of 1,024 possible input strings stored simultaneously in ten qubits. What happens when you reach the end? The answer will be a superposition of all 1,024 solution strings, which are not labelled in any way. Reading one off will not tell you which input led to it, and all the other answers will be destroyed. It's a big stumbling block.

When Deutsch discovered massive parallelism, he realized that it was, essentially, an excellent and powerful solution in search of a suitable problem. Unable to do much with it as it was, he pursued that line of logic and started a hunt for problems that a quantum computer *could* solve. Since then, many others have joined that search, and have come up with a handful of examples. Typically, these problems succeed in illustrating the enormous power of quantum computing, but they are highly contrived. It seemed at the time that quantum computers would be unable to do many things, even though the few things they could do were astonishing. The trick is to define problems that do not require specific solutions of mathematical equations for specific inputs. That kind of information, which is the bread and butter of conventional computers, would clearly tend to get wiped out as soon as you went looking for your answer. Instead, the questions should be specifically tailored to involve generic traits shared by all the solutions. Are they all even numbers? Are they all positive? When plugged into a formula, do they give different answers, or do they all yield the same result?

Since the schemes that have been proposed to date for quantum computation are difficult to describe, I will resort to an analogy instead, with the caveat that it is just that, and not an example of quantum computing. Imagine that you have 1,024 coins each weighing ten grams, except for one counterfeit, which weighs nine. If the coins are all mixed up, and reading a weight on an accurate scale takes a minute, how much time do you need to find the ringer? A straightforward, step-by-step approach, the analogue of *serial computing*, might require anything up to 1,023 minutes – a wearying, tedious seventeen hours, unless you happen to get lucky. If you could enlist ten friends with ten scales you could divide the pile into ten batches, perform the analogue of *parallel computing*, and finish the job in a tenth of the time, a more manageable hour and three-quarters. But if you are really clever, you can use the analogue of *massive parallelism* and do the job by yourself in ten minutes! How? Divide the pile in half, and determine which half is exactly one gram short. Divide that one in half again, and repeat. Keep dividing the pile in half, pick the lighter half each time, and continue. Since 1,024 divided in half ten times equals one, you identify the interloper in a mere ten minutes.

While this little story does not describe quantum computing itself, it does catch some of the right flavour. First of all, the problem is highly contrived. How come there is only one counterfeit, you might ask, and not an unknown number? How come you know that the bad one weighs less, not more than the others? How come the coins all weigh exactly the same? The problem is specially rigged to take best advantage of Shannon's recommended strategy for playing Twenty Questions: ask questions for which the answers 'Yes' and 'No' are about equally likely. The improvement in running time over the straightforward method is startling: no less than a factor of a hundred. Because the ten in the equation $2^{10} = 1,024$ appears in the exponent, massive parallelism is said to entail exponential savings in running time. Most importantly, the clever method does not emphasize particular answers to individual questions, such as: how much does each coin weigh? Instead, it compares aggregate properties: how does this batch of individuals compare to that batch? Just like digits in a quantum computer, the coins are not manipulated one at a time, but many at once.

In 1994, nine years after Deutsch's discovery of massive parallelism, a dramatic breakthrough occurred when Peter Shor at the AT&T Laboratories in New Jersey found its first truly practical application. Even though quantum computers did not exist at the time, and are barely in their infancy today, this event ensured that they would henceforth be taken seriously. Shor's algorithm is a method of dissecting large numbers into prime factors – like figuring out that 164,052 can be written as the product of eight primes, namely $2 \times 2 \times 3 \times 3 \times 3 \times 7 \times 7 \times 31$. This accomplishment, by itself, sounds as artificial as weighing coins, but it turns out that it isn't. It just so happens that the ability to factor large numbers into primes, without too great an investment in time and effort, lies at the heart of cryptography and code-breaking – subjects of burning interest to bankers, secret-service agents, and web merchants. With Shor's algorithm, quantum computing left the ivied halls of universities and think tanks and leaped into the pressured world of high finance and state secrecy. By 1999 proposals had surfaced for wiring the financial district of London, and the government agencies around Washington, with quantum-cryptographic networks that guarantee unbreakable communications. Quantum computing seems destined for a brilliant future.

But what of the present? Can quantum computers really be built? Rolf Landauer and other thoughtful observers were not so sure. Their qualms arose from serious concerns, not mere lack of vision. They knew that quantum computing really does make unprecedented demands on technology and, in turn, that simple trust in technology is never enough. In fact, the very properties that make it so incredibly powerful also render it exceptionally vulnerable. For at the core of quantum computing lies the phenomenon of entanglement – the notion that you can't describe the state of one particle without at the same time describing that of another particle. Entanglement, which arises when two particles interact with each other, allows many qubits to be manipulated together as a unit, because they are inextricably linked together; but since they can interact at a large distance without colliding as violently as billiard balls, the linkage between them can be established very easily – and that might cause trouble.

Suppose an atom with a purposely inscribed qubit encounters some quantum object that is part of the environment, of the physical apparatus that constitutes the machine, or of a cosmic ray that just happened to streak by. If this were to happen, our qubit could become entangled accidentally, and its fate would then be linked to the outside world. If the external member in this unintended partnership subsequently happens to collide with a macroscopic object that could, in principle, allow a measurement, or determination of its value, the entire string of entangled qubits would collapse to a string of random classical bits, and lose its value for the quantum computation. Quantum computing thus requires a degree of quarantine that makes that used for stored smallpox viruses look like a sieve.

A decade ago this hypersensitivity looked as though it might spoil the entire enterprise. However, where there's a will there's a way. Instead of wringing their hands over the possibility of errors, the pioneers of quantum computing accepted their inevitability, and proceeded to invent the art of error correction. Since the capacities of atom-size computers are so huge, they reasoned, perhaps redundancy can come to the rescue. Errors, after all, can be rectified by duplicating, or triplicating, or n-tuplicating each step, and comparing results periodically to weed out mistakes – or even made harmless by so-called 'fault-tolerant' techniques of computing that are not thrown off by occasional small glitches. These efforts have paid off. In a few years the field of error correction has matured to the point that Landauer's fear of the catastrophic effects of excessive sensitivity has been largely allayed.

In spite of serious practical difficulties of construction, there has been progress in the development of actual quantum computers. By 2002 a machine had been built to factor the number 15 into its prime factors 3 and 5. Not enough to secure a bank account, but a successful baby step! A valuable theoretical insight simplifies the architecture of future devices. The building blocks of quantum computers are quantum-mechanical systems that can assume two different states, represented by 0 and 1. It turns out that all two-level systems are described by exactly the same mathematical formulas, so the fundamental theory of operation of a quantum computer does not have to be reinvented for every different technique. Another way of saying this is that software and hardware development do not have to wait for each other, but can proceed in parallel.

Finding actual two-two level systems for possible use as elements of a computer is easy. Photons, for example, can travel through a Mach-Zehnder interferometer along two distinguishable paths labelled 0 and 1. Alternatively, they can be polarized horizontally or vertically and associated with 0 and 1 respectively. Neutrons, too, have been chased through Mach-Zehnder interferometers carved out of single crystals of silicon. Electrons can be lined up with their internal magnets lined up along, or opposed to, the direction of an external magnetic field. Atoms and molecules may have two different allowed energy levels which are, in turn, called 0 and 1. Currents in macroscopic loops of wire may flow clockwise, anticlockwise, or, amazingly, in both directions at once in a quantum-mechanical superposition state. These and countless other large and small systems obey the weird laws that govern qubits. All of them are fair game for an enterprising quantum computer designer.

Once the fundamental cells for storing qubits have been identified, the hard labour of actual construction begins. Provisions must be made for storing input data, either by electrical or optical access to the qubits, for entanglement and subsequent manipulation corresponding to a calculation, and for extracting output information; and all along the qubits must remain sufficiently isolated from their surroundings to avoid spurious entanglements. It's a daunting challenge.

One possible design for quantum computers utilizes trapped ions, atoms that have become electrically charged by losing an electron and are caught in containers as small as a grain of salt by electrical and magnetic forces applied from the outside. If the quantum jumps of their remaining electrons happen to correspond to frequencies of visible light, one can actually see these ions through a microscope – minuscule stars glowing in the darkness of their cage. If half a dozen of them are positioned in a row like a chorus line, the resulting structure represents a six-qubit quantum register. As long as the ions are securely held, and do not approach any walls, and as long as their container is purged of stray air molecules, they are well isolated; but they still need to be entangled with each other, and addressed individually for purposes of data input and output. The former is achieved by gently shaking the entire chorus line: in order to participate in a joint dance they must communicate with each other electrically, and such an exchange entails entanglement. Other chores, such as preparation and readout, are accomplished with the help of finely collimated and tuned laser beams that move electrons around in the outer shells of the ions like so many beads on an abacus. Ion traps are pretty devices, and hold great potential for quantum computing, but they are a long way from industrial application.

A completely different technique builds on decades of experience with magnetic-resonance imaging, or MRI. (The name is a neologism for the phrase 'nuclear magnetic resonance', or NMR, which was deemed to sound too threatening for the general public.) In NMR the tiny magnets associated with atomic nuclei line up along, or opposed to, external magnetic fields to make two-state systems. They are manipulated by radio signals that can flip the nuclei around like so many hamburgers, provided the frequency is adjusted just right to resonate with the nuclear beat. The nuclei in question are typically those of one specific atom in a molecule – say, hydrogen – held in place by normal chemical forces. The difference between ion traps and NMR is dramatic. The first technique relies on single atoms, or a few at best, kept in isolation from the rest of the world. NMR, on the other hand, uses the molecules in a flask of liquid – around 10^{23} of them, all acting in unison. These nuclei are isolated too, but the quarantine is natural, and provided for each nucleus by its proper molecule. Since the molecules are identical, their internal states are identical too. This implies that we are not dealing with 10^{23} different qubits, but only with one or two – as many as a single molecule can hold. For purposes of error correction, NMR is a dream come true. Imagine how many atomic errors a test tube full of liquid can tolerate before their signal strength becomes comparable to that of the bulk!

Ion traps, NMR, and a dozen other techniques are being studied in the quantum computer labs of the world. In spite of much progress, the winning design may not resemble any of those that exist today.

Rolf Landauer's doubts about the feasibility of quantum computing turn out to have been exaggerated; but he also raised a much more fundamental concern with his remark that eventually computers will affect the very laws of physics on which they depend. The idea, which like so many other good ones within the discipline owes a lot to Richard Feynman, is this. The aim of physics is, as far as possible, to map out the material world in terms of mathematical formulas and complex equations that can then be solved numerically. Cutting out the human middleman, as it were, the scheme is to map the material world onto a computer program in what is known as a simulation, or model. That much of modern physics has taken on the character of model-building, rather than establishing the true laws of nature, is illustrated by the fact that the current theory of elementary particles is called the Standard Model, whereas Newton's theory of the mechanical world still carries the proud title Newton's laws.

The complexity, and indeed the very nature, of a model is limited by the capabilities of the computer that grinds it out – something that becomes abundantly clear when you consider mapping a quantum-mechanical system. The first thing you would need is a ready source of random numbers, because randomness lies at the very heart of atomic behaviour; but your computer cannot furnish them! No program can simulate random numbers, because by definition they obey no rules, and conventional computers operate exclusively by rule. So right from the start you have to reach outside your model to find a set of random numbers somewhere in nature – perhaps in a real quantum system, such as a beam splitter.

This little quandary is, in fact, the tip of an iceberg, and one that hints at layers of difficulty that lie beneath the surface. Since quantum mechanics describes not only what is, but also the huge number of things that might potentially be, it is also too rich to be modelled by a conventional computer. Consider a little group of 100 atomic magnets lined up like pennies in a row, some facing up, others down. Classically this system is described by a single string one hundred digits long. Quantum-mechanically, on the other hand, it corresponds to a hundred qubits, which collectively have 2^{100} or about 10^{30} states. No supercomputer built today or in the next century can handle that volume of information; but quantum computers, which operate with qubits rather than bits, *do* have sufficient power to model this or any other quantum-mechanical system; and since we live in a quantum-mechanical world, this means that only a quantum computer can produce a reliable model of the world.

There is something disturbing in this conclusion. I remember a physics course I took long ago on the subject of molecular spectra. Molecules of gases like nitrogen dioxide consist of several atoms, so the spectrum of colours emitted when the gas is heated can become exceedingly complicated. In turn, the solution of Schrödinger's equation, which describes the molecule completely, became exceedingly difficult. Since digital computers had not yet become common at the time, we students became more and more frustrated in our struggle to keep up with the maths. Finally the professor relented and said, half in jest: 'In the end, we cannot keep up by hand, and must resort to an analogue computer to solve Schrödinger's equation for us; and the best analogue computer of all is the nitrogen dioxide molecule itself. So we just read off the solutions it finds by measuring the wavelengths of the spectral lines.'

A quantum computer resembles its digital counterpart in its input and output, but in its internal workings it uses natural objects such as ions and molecules to solve the Schrödinger equation for us. This scheme resembles my professor's suggestion of switching to an analogue computer, but differs in one important respect. The system to be modelled is not the same as the system that is used as a computer, but may be quite different. For example, a nanometre-size nitrogen dioxide molecule may be simulated on an NMR computer that uses femtometre-size hydrogen nuclei in a cup of liquid – on a scale a million times smaller. In a way, then, quantum computing comes down to simulating complex, mysterious parts of nature by comparing them to simpler but equally mysterious parts of nature; but both the large molecule and the tiny nucleus are governed by the same rules of quantum mechanics.

This is not what I thought physics was about when I started out: I learned that the idea is to explain nature in terms of clearly understood mathematical laws; but perhaps comparisons are the best we can hope for. Finding relationships between apparently disparate phenomena – nitrogen dioxide molecules and hydrogen nuclei – and thus simplifying our world picture may be all that nature allows; but that's quite a lot, actually. It is, in fact, how Newton started out when he compared the motion of the Moon to the fall of an apple. However, Newton never discovered the real nature of gravity. If we try to understand nature by simulating quantum systems on a quantum computer, we may never learn to understand either one.

23 Black Holes

Where information goes to hide

Paradox is the sharpest scalpel in the satchel of science. Nothing concentrates the mind as effectively, regardless of whether it pits two competing theories against each other, or theory against observation, or a compelling mathematical deduction against ordinary common sense. 'We will never understand anything until we have found some contradictions,' exclaimed Niels Bohr; and few phenomena in physics have generated more paradoxes than those fabulous denizens of outer space, black holes – the more complex of which have also, through their resolution, served to reveal much about the concept of information.

Let's start with a simple example, one already apparent in the definition of black holes as small, improbably dense objects that are too heavy for anything, including light, to escape their embrace. The suggestion that even light should fall victim to gravity goes against every direct experience we have of it. Nevertheless, the phenomenon is both a firm prediction of Einstein's general theory of relativity and a stubborn fact that has been corroborated by observations going back to 1919. *Everything* falls under gravity: baseballs, buttered toast, atoms and photons. So there's nothing contradictory about a celestial object that pulls every material particle and every photon ejected from it back to its surface.

A second paradox lies in the fact that, while we can't see black holes, astronomers claim to have plausible evidence for their existence, not only in our own galaxy but in neighbouring ones as well. The solution rests on the commonplace remark that indirect evidence can be as strong as direct evidence. You don't have to open a cherry to know there's a stone in the middle! Black holes reveal their presence by the gas and dust particles that revolve around them in a disc, and sometimes spiral into them, heating up to incandescence in the process and sending out torrents of radiation in the form of visible light as well as X-rays and radio waves. Such clues have convinced all but the most sceptical astronomers that black holes actually exist.

While the above two paradoxes are easily worked round, there are far trickier ones out there. In 1971 John Wheeler, the guru of information science who brought black holes into the mainstream of physics and astronomy, pointed out to his graduate student Jacob Bekenstein that black holes seem to flout the Second Law of Thermodynamics. Imagine mixing a cup of hot tea with a cup of cold water – creating disorder and entropy in the process – and then pouring the tepid mixture into a black hole. The newly generated entropy disappears from the universe permanently. There is no way to get it back, or to compute its value once it has vanished; indeed, without any apparent mechanism to balance things out, it seems to have done the impossible and decreased. Wheeler, always ready with an apt phrase, called this destruction of the evidence of illegal activity a 'perfect crime'.

Bekenstein, less brazen than his mentor, refused to believe that such a sturdy cornerstone of physics could be set aside so easily, and set about looking for an explanation. He quickly succeeded. Starting with a firm belief in the Second Law, he argued that a black hole must have an entropy of its own, which increases when things like weak tea are thrown into it. Based on his understanding of the behaviour of black holes, he guessed that the physical property of a black hole that would reveal this hidden rise in entropy to the outside world would be its surface area. Although this sounds like an off-the-wall suggestion, it turns out that the size, and hence the surface area, of a black hole is related to its mass: the heavier, the bigger, just like our own bodies. Via Einstein's universal $E = mc^2$, entropy is thus related to energy, as it was when

Rudolf Clausius created the concept as a counterpoint to energy.

Shortly before Bekenstein made his outrageous suggestion, it had become clear that, whenever black holes interact with matter and with each other, their total surface area does tend to increase. Based on this mathematical consequence of the theory of relativity, he therefore conjectured a new law of nature, the 'Generalized Second Law': the sum of black-hole entropy and ordinary entropy outside the black hole never decreases. Following the lead of the thermodynamicists of the nineteenth century, Bekenstein then checked his hypothesis against a number of special cases, and found that it held.

The reaction of the physics community ranged from incredulity to indifference. If the association of entropy with surface area were not weird enough, there was simply no precedent for mixing general relativity – the theory of gravity that deals with black holes – with thermodynamics, the science of tea kettles and steam engines. Beyond indifference, there was active hostility as well. The leader of the opposition was Stephen Hawking, who argued that since entropy is heat energy divided by temperature, a black hole with entropy must have a temperature, and if it has a temperature, it must radiate. All warm bodies radiate – visible light when incandescent, infrared or heat rays when cool. But black holes, by definition, don't radiate. No radiation implies no temperature, which implies no entropy, *ergo* Bekenstein is wrong. All the leading experts, with the notable exception of John Wheeler, agreed with him.

Then a miracle occurred. Like Saul on the road to Damascus, Hawking saw the light and was converted. His status as scientific superstar rests on his discovery in 1974 that black holes, contrary to received wisdom and his own firm expectations, *do* radiate. So they have a temperature after all, and an entropy, and Bekenstein was right all along! To be sure, since nothing can escape the interior of a black hole, Hawking radiation differs in nature from sunlight. In a tour de force of consilience, Hawking melded ideas from general relativity, thermodynamics and quantum mechanics to show how the radiation originates from the intense gravitational field just outside the surface of the black hole. Indeed, Hawking radiation is as integral a component of its parent body as the atmosphere is of our own planet. Black holes turn out to be cool because Hawking radiation is weak – so unimaginably feeble that it will surely never be detected directly – but its existence is nevertheless now universally accepted. It carries energy as well as entropy away into space, and with them, presumably information.

After the invention of classical entropy in the middle of the nineteenth century, thermodynamics evolved into statistical mechanics – the physics of many constituents – and to the equation on Boltzmann's tombstone. Entropy was derived from the underlying atomic doctrine, and related to the number of ways a roomful of air, a cup of tea or any complex system can be rearranged without changing its macroscopic, measurable aspects. After Hawking's discovery, astrophysicists naturally set off along the same road. They accepted Bekenstein's formula for black-hole entropy in terms of surface area, and enthroned it alongside the old definition of entropy as heat over temperature. Now the hunt was on for a re-interpretation of black-hole entropy in terms of the number of ways of rearranging its microscopic components. However, they quickly ran into a seemingly insurmountable obstacle: how do you count the number of ways a black hole can rearrange itself if you don't know anything about its interior?

Black holes distinguish themselves from all other systems by hiding their complexity under an impenetrable veil. Undaunted, physicists proposed a number of detours around this difficulty. Bekenstein, for example, put forward the argument that what matters is not so much what a black hole *is* as how a black hole *becomes* what it is. He proposed, therefore, to count the number of quantum configurations of any matter that could collapse to form the black hole; but a careful accounting of the large amount of fresh entropy added by throwing additional matter down the hole, minus the meagre flow of entropy carried off by Hawking radiation, once this theoretical act of creation had been performed, didn't add up very well.

Proponents of string theory achieved a more successful derivation of Bekenstein's formula from statistical mechanics in 1996. Using an explicit quantum theory of matter and forces, including, for the first time, the force of gravity, they counted the number of ways a certain 'mathematical object' in their theory could be rearranged with impunity. The logarithm of the result, they triumphantly concluded, reproduced the correct Bekenstein entropy. Unfortunately, however, though the toy model they were investigating has many of the properties of a black hole, it doesn't have all of them. In particular, the model is frozen at zero temperature, so it certainly does not describe a 'normal' black hole emitting Hawking radiation. Since then, more complicated string-theoretical models, called *Gedanken* (German for thought) black holes because they exist on paper but not in the real world, have been analysed with the same encouraging result. Nevertheless, Bekenstein recommends caution: 'If you have lost a nickel, and somebody has found a nickel, it does not prove it is the same nickel. Are [string-theory] black holes and traditional black holes the same thing?'

Until black-hole entropy can be reliably reduced to statistical mechanics, thermodynamics returns to the status it had in the interval between Clausius's formulation of the Second Law in terms of entropy, and Boltzmann's later explanation of entropy as the log of the number of ways. Thermodynamics is once more an independent, universal, physical theory. Most physicists take the view that in everyday life, far from the black holes of outer space, the Second Law is a consequence of more fundamental mechanical and statistical assumptions; but from a cosmic point of view, the entropy of black holes has restored the primacy of thermodynamics over statistical mechanics. In a curious parallel, Einstein's special theory of relativity had a similar effect on the *First* Law of Thermodynamics, the principle of energy conservation. In ordinary contexts, conservation of energy follows mathematically from the well-established laws of classical mechanics, electromagnetism and quantum mechanics; but when Einstein introduced his new formulation of the laws of mechanics, he found that the principle of conservation of energy would retain its validity *only* if the unexpected formula $E = mc^2$ were also adopted. Thus, far from following automatically from the theory of relativity, the law of conservation of energy suddenly regained its primacy. Einstein's two theories of relativity have had the unanticipated effect of generalizing Rudolf Clausius's universal laws of thermodynamics. In modern dress they read: 1) The energy of the universe, including the rest energy of massive bodies, is a constant. 2) The entropy of the universe, including the entropy of black holes, tends to increase.

The most perplexing contradiction concerning black holes is Stephen Hawking's infamous Black Hole Information Paradox. The problem is quantum-mechanical. Imagine a quantum system – say, two hydrogen atoms cavorting in a box. They can be completely described by a wave function, which encodes all the information we will ever be able to find about them. By manipulating this in an appropriate manner, we can learn some things about the atoms with certainty – their number, for example – and others with varying levels of probability. As long as we don't reach into the box to disturb the system, the wave function will evolve smoothly and absolutely predictably, because it obeys an equation that we understand perfectly and can even solve explicitly for simple systems. The sum total of all the information encoded in the wave function remains constant. *Information is conserved in a natural process described by quantum mechanics.*

Look what happens, though, if you take those two atoms and throw them into a black hole. The information we had about them disappears. To be sure, the black hole has become a little heavier in the process, but don't worry about that; Hawking radiation is escaping from it, and soon it is back to its former state – precisely its former state. The net effect of the experiment – a refinement of Wheeler's tepid-tea-disposal project – is that the black hole acts as a kind of catalyst to convert a system described by a wave function into another system that includes thermal radiation. Unfortunately thermal radiation encodes much less information than a quantum state. All the complications of the two hydrogen atoms are replaced by one single number: the temperature of the radiation

209

from the black hole. Information, in other words, is lost!

By losing information, a black hole achieves something that is forbidden by the rules of quantum mechanics. Much head-scratching has resulted from this conundrum. Some physicists take the radical view that black holes are in some way the universe's paper-shredders and actually destroy information. If that con-tradicts quantum mechanics, maybe the rules have to be amended to apply to the extreme conditions of pressure and density of matter, and the attendant tortured crumpling of space-time, that prevail inside black holes. Others think that maybe information is quietly carried away by Hawking radiation, encoded in ways we haven't figured out yet. Or perhaps the information is merely hiding, only to be released in a burst at the end of the black hole's life when it has finally exhausted its energy reserves. A still stranger hypothesis has the information sucked out of the interior of the black hole through a wormhole, a sort of tunnel to another uni-verse; but, as Hawking himself wrote at the end of an encyclopedia essay on black holes, at this point '... we run the danger of trespassing beyond the borders of present knowledge and under-standing'.

Talk of the information contained in black holes sounds highly metaphysical, but it turns out that it can actually throw light on some mundane, everyday problems. By tying together the requirements of the most fundamental laws of physics, black-hole theories provide clues for such questions as: 'What are the limits set by nature on the growth of information technology?' In particular, black holes can help us to uncover the limits for computer mem-ories; for as they continue to grow in density and shrink in size, it is worth asking exactly how much information the laws of nature allow them to hold. An estimate of the number of bits that can be stored in a solid chunk of material, extrapolated from currently available technology, is about one per atom, or roughly 10^{20} in a cubic centimetre. If you stretch your imagination to shrink this cube into a black hole, and use the known formula for its entropy to estimate the theoretical maximum amount of information it can store, you get an ultimate limit of about 10^{65} bits per cc, a bound which no memory can exceed. The gap of forty-five orders of magnitude between what might be technically achievable, and what fundamental theory allows, seems so ridiculously large that it renders the black-hole argument irrelevant.

More stringent bounds can be found, however. Bekenstein, for example, suggests that instead of letting the cube of matter shrink into a black hole, one should just drop it gently into a pre-existing black hole. The radius of the host will increase, and so will its entropy – by a known amount. Using this argument, a different theoretical bound of no more than 10^{38} bits per cc emerges – a gigantic twenty-seven orders of magnitude closer to achievability than the previous estimate. Remarkably, Bekenstein's method uses the black hole only as a kind of conceptual tool, and ends up with a formula that doesn't even refer to gravity. Accordingly it looks more down-to-earth than the first calculation, although it is still eighteen orders of magnitude away from a realistic technological bound. The point of such arguments is not to predict what the future holds for computers, but to try to flesh out the meaning of information.

One of the dominant themes in physics today is the search for a reconciliation of general relativity – the description of space, time and matter in the large – with quantum mechanics – the language of the atom. The former is cast in terms of geometry and the gravitational field, while the latter makes use of wave functions. In order to express their incompatibility, it is first necessary to find common terms of reference. The Black Hole Information Paradox demonstrates that information is just such a common concept. It enters general relativity via Bekenstein's entropy, and quantum mechanics via the probability interpretation of the wave function. By its generality, information thus finds itself in the centre of what Hawking calls '. . . one of the major questions in theoretical physics today'. If paradox is a harbinger of progress, as Bohr maintained, the concept of information has a brilliant future.

Work in Progress

Information theory beyond Shannon

Consider the bookie in an alley behind a racetrack, furtively nego-tiating with a stable hand for a hot tip. How much should he offer for it? What's the asking price? What, in loftier language, is the market value of the information? The only currency the bookie knows is cash, and somehow, relying on instinct, experience and street savvy, he comes up with a figure and makes the deal.

Although he has never heard of him, our bookie leaves Shannon in the dust. Not only is he using different units for measuring information – bucks instead of bits – but he is most definitely interested in its meaning and quality – not just its quantity and speed. Can we learn something from him, the way the math-ematicians of the seventeenth century who invented the theory of probability learned from dice and cards?

Questioning Shannon's authority borders on blasphemy. He is the pioneer of the information age, the Pythagoras of information science. Indeed, his influence extends way beyond mere science and engineering: the technology he helped to create is shaping our world; but no prophet remains unchallenged for ever. Indeed, the phenomenal success of information technology over the last fifty years requires that its foundations be re-examined, and its dogmas questioned. Even the Pythagorean theorem itself did not emerge as a cornerstone of geometry until it was placed on the solid footing of Euclid's axioms 200 years after its discovery. The beginning of the twenty-first century seems like an appropriate time to return to basics by asking: what assumptions underlie

Shannon's theory? Are bits unique, or are there alternatives? And if it turns out that there are other ways of measuring information, how do you choose between them?

The realization that there are rivals to Shannon's logarithmic measure of information is not confined to the back alleys of racetracks, but has penetrated academia as well. As early as 1925, a quarter of a century before Shannon's paper, the British biostatistician Sir Ronald Aylmer Fisher proposed a way of measuring the information content of continuous, rather than digital data. The tool he used was not the binary code of the computer, but the statistical distribution of classical probability theory. (An example of a distribution function might answer a question like: 'How many fishes in this pond are between 5.2 and 5.3 centimetres long?') Anticipating Shannon, Fisher wrote: 'As a mathematical quantity information is strikingly similar to *entropy* in the mathematical theory of thermo-dynamics.' Much later, in 1980, William Wootters, who helped to coin the world 'qubit', wrote a PhD dissertation under the guidance of John Wheeler in which he pointed out the similarity between the theory of Fisher information, as it came to be called, and quantum mechanics. They share a common mathematical formalism, and both relate probabilities to the squares of continuous functions. (In quantum mechanics, a typical question might be: 'What's the probability of finding the electron in an orbit between 5.2 and 5.3 nanometres from the nucleus?') More recently, Fisher information has been advanced as a fundamental concept for unifying *all* of physics, but the proposal has not found many supporters.

However, picking out one specific new scheme or another for measuring information is not the only approach: a more radical one would be to follow Euclid's example by searching for fundamental axioms. Indeed, beginning with Shannon himself, many scientists have striven to derive his equations from general assumptions about the way any information measure should work; but, as with Bekenstein's formula, the very fact that different sets of assumptions end up leading to the same conclusions should raise the suspicion that the proposed axioms may be unintentionally biased towards Shannon's results. What's needed is a fresh, unprejudiced examination of the problem of quantifying information – something that Jan Kåhre, a recent challenger to the Shannon hegemony, has provided.

Kåhre hails from Åland, an autonomous group of islands in the Baltic Sea between Finland and Sweden and inherited his independence as a birthright. The Åland Islands constitute a province of Finland, but because they are unilingually Swedish, they have, over the centuries, acquired a high degree of self-governance recognized by their own country as well as by the European Union. They are officially neutral in armed conflicts, and totally demilitarized. Self-determination and self-confidence would seem to be values that Ålanders absorb with their mothers' milk.

Kåhre is an engineer without a formal academic affiliation. Inspired by practical problems of automated control of industrial processes, he became acquainted with information theory but quickly grew disillusioned with the lack of focus he found in the voluminous literature written on the subject. Even though other information measures besides Shannon's were mentioned here and there, he missed a way of comparing them, and a guiding principle to unite them. Independent-minded and undaunted by his outsider status, Kåhre set out to build a better foundation for information science.

In 2002 he collected his researches in *The Mathematical Theory of Information*, a title purposely chosen to echo that of Shannon's own book *The Mathematical Theory of Communication*. In five hundred engaging, erudite and aggressively iconoclastic pages Kåhre achieves two principal goals: to demonstrate convincingly that Shannon's is only one of an infinite number of ways of measuring information, and to propose a fundamental law, a basic principle to which *all* legitimate information measures must conform.

The bookie's dilemma is Kåhre's first illustration. Evaluating the gambling tip by counting bits is a useless exercise. It's a question of utility. What the bookie needs is a measure of quality before he can figure out a cash value on the information he wants to buy. The appropriate concept turns out to be 'reliability', the probability that a message is correct. For the bookie, a tip that is 100 per cent sure would be worth his expected betting profits in cash, but if it is only 50 per cent certain it is worth nothing – no better than a coin toss. Curiously, if the stable hand's tip is sure to be wrong, it is worth as much as if it were reliable. Consistent lying, as we saw with interferometers, is a back-handed form of telling the truth. Kåhre shows how the bookie, combining his estimate of reliability with his understanding of betting odds, can easily devise a way of assigning a cash value to information, in bucks rather than bits.

Since Shannon restricted himself to information transmission with 100 per cent certainty, he did not need to figure reliability into his equations. More sophisticated branches of information technology that have grown up since then cannot afford this luxury. In the modern discipline of *pattern classification*, for example, accuracy in discriminating between two patterns, such as a zero or a one, is fundamental. In pattern classification accuracy is measured in the number of correct classifications, which are called 'hits'. Kåhre adopts this nomenclature, and concludes that whereas bits and entropy are quantitative information measures, hits and reliability measure quality. 'Hits before bits' is the enthusiastic Wheeleresque war cry adopted by Jan Hajek, a Dutch connoisseur of information theory.

In addition to bits, bucks, hits, reliability and Fisher information, Kåhre also covers a way of weighing information that is used by scholars of the allied disciplines of game theory and economic decision analysis. There, as in the bookie's case, the problem is to find the value of a piece of information in helping a player to win or a capitalist to profit. The mathematical expression of this value is called the 'utility', a numerical measure favoured by the Hungarian-American mathematician John von Neumann. In his honour, Kåhre coins the term 'Nut', meaning 'von Neumann utility'. This together with those methods mentioned above are, however, just a handful of an infinite number of possible information measures described in his book. Each is appropriate in its own sphere, and none is more fundamental than others. Bits take pride of place for both historical and technical reasons, but contrary to received wisdom he shows that they are by no means unique.

Kåhre achieved his second aim – to find a common denominator among the different ways of measuring information – in a fundamental axiom he calls the *Law of Diminishing Information*. It is, in his words, the pruning knife of information theory: if a mathematical function obeys the law, it is acceptable as a possible measure of information; conversely, if a function violates the law, it must be rejected as unsuitable.

The Law of Diminishing Information – LDI for short – concerns transmission of information through two channels in a row, such as two telephone wires spliced together, or a message delivered to a receiver who in turn feeds it into a computer, where it can be manipulated and then printed out for a second receiver. Briefly stated, LDI stipulates that compared to direct reception, an intermediary – the second channel in the chain – can either leave the amount of information unaffected, or it can diminish the information. What it cannot do is add more information.

From one point of view this rule is both obvious and familiar. Everybody has played the game of Chinese Whispers, or Telephone, in which one child chooses a sentence, whispers it into the ear of a second child sitting in a row, who in turn whispers it to the third child, and so on down the line. The message that emerges at the end of the line is usually hilariously garbled and seemingly unrelated to the original sentence. It's a classic example of the corrupting effect of noise on information transmission, and illustrates LDI on a primitive level.

However, the law also has a much more profound aspect, which Kåhre introduces by reference to a Chinese newspaper. Imagine an Englishman without knowledge of the Chinese language who decides to evaluate the paper. Since the foreign characters have no meaning for him, he concludes that the information content of the paper is zero; but then he meets an interpreter who translates the paper. Now, using the same criteria as before, the Englishman assigns a high information content to the paper. The intermediary channel, in other words, has managed to raise the information content of the original message. The fact that this scenario violates the LDI signals a problem.

We have met this issue of subjectivity before. In trying to understand the nature of information, we asked: how much information does the sequence 14159265 ... contain? To some it is random gibberish, devoid of information; but to others it is the fractional part of the constant π, and represents an important piece of mathematical information. Any information measure that depends so strongly on the prior knowledge of the recipient is too subjective to be of use, and should be discarded. Thus the Englishman's judgement of the Chinese paper must be considered faulty, and cannot be admitted as a legitimate measure of information – as its failure to satisfy the LDI suggests. Shannon's logarithm, on the other hand, passes the LDI test with flying colours, largely because it purposely eschews the question of meaning.

With the LDI as cornerstone, Kåhre develops his mathematical theory of information. It clarifies and relates different measures of information, rejects some that have been proposed, invents entire families of new measures that qualify, and proves general theorems about the behaviour of information. Kåhre worries about the wisdom of basing his theory on a postulate as negative as the LDI – an intermediate channel shall not increase the information. As a physicist, I tend to be more sanguine, remembering that the entire edifice of thermodynamics is founded on two simple statements, of which the second, the law that entropy tends to increase while order decreases, sounds just as negative, and furthermore resembles the Law of Diminishing Information to a remarkable degree.

Kåhre's theory is not the bolt of lightning Shannon's was, but a consolidation of divergent directions of research and a liberalization of thinking. It helps to free information theory from the straitjacket of Shannon theory into which it has been squeezed – not by Shannon himself, but, as happens so often with great prophets, by his followers. With respect to classical information, the progress beyond bit-counting requires a measure that depends to some extent on the meaning of the message. As every bookie knows instinctively, a number such as *reliability* – a qualitative rather than quantitative measure – is needed to make the valuation of information practically useful. Bucks, hits and Nuts all incorporate some measure of reliability, and thus represent a tiny step in the direction of the broad, literary definition of the word 'information' that we began by looking for.

Kåhre deals principally with classical information, and has little to say about the strange world of the quantum; and yet, his survey of acceptable measures of information is sufficiently thorough to include the classical version of a new measure of quantum information that has recently been proposed. To learn about that one we have to leave the charming Åland Islands and return once more to Vienna, the birthplace of Boltzmann and hence of information as a physical concept.

25 Zeilinger's Principle

Information at the root of reality

The spacious Institute of Experimental Physics of the University of Vienna, where Anton Zeilinger has his laboratories, is located at 5 Boltzmanngasse, across the street from the Hotel Boltzmann. When I visited him, I found the pavement outside had been torn up after the current drawn by Zeilinger's powerful lasers had finally overtaxed the venerable institute's circuits, so that they needed rewiring. At the head of the street, where it branches off from the principal artery, a blue enamel sign explains the origin of its name: 'Ludwig Boltzmann (1844–1906) physicist and philosopher.'

The description of Boltzmann as philosopher comes as a surprise to the international visitors who flock to Zeilinger's labs. Not even the official guidebook for the celebrity section of the Central Cemetery, where Boltzmann is buried, claims that he was anything but a physicist. Nevertheless, the street sign contains an element of truth that the functionaries who erected it may not have realized. Regardless of whether the cause is to be found in the Austrian educational system or in a penchant for contemplation in the Austrian national character, it seems that the grand tradition of Austrian physics does indeed include a heavy dose of philosophy. Its world view contrasts sharply with the equally successful, but much more pragmatic American approach. Zeilinger's most recent work can be appreciated more fully when it is placed in the context of the intellectual tradition to which he is a proud heir.

Boltzmann's philosophical frame of mind was evident from boyhood onwards and underlined in later life by the topics he chose for his lectures, which ranged widely over philosophical topics, from those that might be expected to occupy a philosophically inclined physicist, such as: 'On the Question of the Objective Existence of Processes in Inanimate Nature', to polemically charged outbursts such as: 'Proof that Schopenhauer Is a Mindless, Ignorant Philosophaster Who Peddles Nonsense and, Through his Empty Verbiage, Is a Thoroughly and Forever Degenerating Influence on Others'. Above all, though, Boltzmann's philosophical approach to physics is manifested in his lifelong efforts to uncover the *meaning* of entropy. The original definition as 'heat divided by temperature', proposed by Rudolf Clausius, was perfectly acceptable, and, what's more, enormously useful for experimentalists, but it shed no light on the concept's real significance. It was Boltzmann's determination to address this embarrassing question that qualified him as a philosopher, while his answer elevated him to the first rank of physicists. How profoundly his attitude contrasted with that of Claude Shannon.

A colleague of Boltzmann's at the University of Vienna, and a stubborn adversary when it came to accepting the reality of atoms, was Ernst Mach. Originally an experimentalist, Mach turned increasingly to philosophy in his later years. His careful analysis of the meaning of space-time prompted Einstein to claim: 'The type of critical thinking which was required for the discovery of this central point was decisively furthered, in my case, especially by reading of David Hume's and Ernst Mach's philosophical writings.' No more persuasive argument could be advanced for legitimizing the role of philosophy in physics.

Another Viennese professor of physics, Erwin Schrödinger, the author of *What is Life?*, also had an intensely philosophical outlook on the world. As soon as Heisenberg published a quantum theory in terms of the strange, forbidding spreadsheets called matrices, Schrödinger set out to create something more intuitively acceptable than a mere recipe for making predictions – no matter how successful it might turn out to be. 'Physics does not consist only of atomic research,' he wrote, 'science does not consist only of physics, and life does not consist only of science. The aim of atomic research is to fit our empirical knowledge concerning it into our other thinking.' This opinion was at the core of Schrödinger's humanistic philosophy.

And it doesn't stop there. Even Austrian physicists whose careers took them abroad were not immune to the national infection. Wolfgang Pauli was born in Vienna in 1900, and came under the local scientific tradition's influence by way of his father, who was the professor of physical chemistry at the city's university. Pauli himself studied in Munich, then went on to academic positions in Germany and Switzerland, and later, to a Nobel prize. He was perhaps the most accomplished mathematician among the principal architects of quantum theory, and so severe a critic of his colleagues that he became known as the 'conscience of physics', not in moral terms, but in the sense of a master craftsman defending the integrity of his trade against the bungling of the less talented. Yet Pauli, that most rigorous of abstract theoreticians, was also above all a philosopher. In a moving obituary, his student Marcus Fierz, himself an eminent physicist, wrote: 'He was a natural philosopher in the classical sense of the phrase – as it applies to Kepler, Galileo, and Newton. Like hardly another of today's physicists, Pauli was imbued with a deep feeling that the scientist's efforts in reading Nature's book would lead man towards recognition of his own image in Nature itself ... For Pauli the basic questions of physics were in their deepest meaning also the questions of human life.'

With this background, is it any wonder that Anton Zeilinger, Professor of Experimental Physics at the University of Vienna, also feels that a philosophical attitude, while not necessary, is nevertheless highly commendable in the pursuit of fundamental physics?

Of course, such an outlook is not enough by itself. To win respect in the profession, a physicist must above all make a lasting contribution to hard-core science, as Boltzmann, Mach, Schrödinger, and Pauli unquestionably did – a test that Zeilinger passes with flying colours. He first came to the attention of the physics community in 1989 when, together with Daniel M. Greenberger of City University of New York and Michael Horne of Stoneham College near Boston, he proposed a surprising and ingenious experiment, now known simply as GHZ, for testing the weirdness of quantum mechanics by using a system of three entangled particles, rather than just the two that date back to Einstein, Podolsky and Rosen. At first this, too, was a *Gedankenexperiment* – a thought experiment – in the venerable tradition of hypothetical exercises that compare quantum-mechanical with classical predictions. However, when the experiment was actually carried out some ten years later, no one was surprised that quantum theory had once more been decisively corroborated. Nevertheless, GHZ is considered a classic, not least because of its conceptual simplicity, and has inspired an enormous amount of additional theoretical and experimental work.

In the 1990s Zeilinger went on to perform spectacular experiments in quantum teleportation – a misnomer in the sense that neither material nor information is actually transported instantaneously through space, but an important step in the experimental investigation of the nature of quantum mechanics, as well as a potentially useful element of the emerging field of quantum information technology. Most recently he and his students have set world records in the mass and complexity of objects that can be shown to display quantum-mechanical wave interference. Where once physicists were astonished to see waviness manifested in 'heavy' objects like neutrons, and complicated systems like atoms, Zeilinger now reveals it in atomically Gargantuan buckyballs, giant molecules with 60 or 70 carbon atoms. Viruses are next.

Today Zeilinger is a busy man. Even as he directs a large group of students and assistants, he flits around the world attending conferences and leading professional seminars. The local press demands interviews on the day's hot topics, demonstration experiments need to be prepared for his large introductory physics lectures at the university, articles wait to be written for technical and popular consumption, colleagues require attention. Zeilinger doesn't have much leisure, the philosopher's most precious commodity, but from the way his eyes light up when he talks about the deep, philosophical problems surrounding the interpretation of quantum mechanics, it is obvious where his heart really lies. His rumpled, avuncular appearance belies a sharp wit, which he is quick to apply to any subject from Austrian politics to modern art; but underneath one senses a single-minded, passionate drive to understand the meaning of existence, which his life's work has convinced him must somehow be related to the enigma of the quantum. Like John Wheeler he wants to know: why the quantum? Like Schrödinger he wants the answer to speak to the broad public; and like Pauli he feels intuitively that the answer, when it comes, will turn out to illuminate human nature itself.

In 1996 Zeilinger undertook a critical review of the literature on interpretations of quantum mechanics, paying special attention to what the masters themselves said in private letters. He reported his findings in a paper entitled 'On the Interpretation and Philosophical Foundations of Quantum Mechanics', and his cautiously expressed conclusion, 'that there might be a problem of proper philosophical foundation in quantum mechanics', led him to wonder what such a foundation might look like. The proliferation of alternative interpretations of quantum theory in the last half-century might be evidence, he suggested, that they might not be as solid as they should be. What's missing, he thought, was a *Grundprinzip* – a basic principle – something clear and simple and firm.

By way of illustration, Zeilinger cited Einstein's principle of special relativity, which holds that the laws of physics should be the same in a stationary laboratory as in a rocket ship moving at a high, constant speed. This principle goes back to Galileo – whose imaginary ship was propelled by sails instead of rockets – and is universally accepted. Even though most of the specific equations of relativity predated Einstein's work, they were not adopted before Einstein, Zeilinger suggests, because they lacked a firm foundation. The principle of relativity finally furnished such a philosophical grounding, and in consequence Einstein's theory was quickly understood and incorporated into the body of physics.

It's a useful analogy, but quantum theory is different, for while no one doubts its basic correctness and phenomenal success, no one understands it, even though, as Einstein wrote to Schrödinger about the majority of physicists: '... the guys resist admitting this (even to themselves).' (Einstein referred to them as *'die Kerle,'* which has pejorative overtones of churlishness missing from the word 'guys'.)

Nonetheless, Zeilinger began his search for that illusive unifying principle, and three years later, in 1999, he thought he had found it. On 5 January he submitted a paper with the bold title 'A Foundational Principle for Quantum Mechanics'. Like the principle of relativity and the law of conservation of energy, it sounds remarkably innocuous:

An elementary system carries one bit of information.

Over Greek salad in one of the picturesque outdoor cafeterias of the University of Vienna he explained it to me.

Its crucial, underlying assumption is Bohr's proclamation that physics in general, and quantum mechanics in particular, do not describe the world itself, only what we are able to say about it. Bohr demands that we wake up from the spell of Democritus – the illusion that we can come to grips with the objective material world, without acknowledging, or even trying to understand, the mediating role of information. We never see a chair, Bohr would say: we receive sense impressions that give us information which our brains somehow process into the idea (an Aristotelian 'Form') of chair. We don't see, or detect, or measure an atom: we gather information about the atom and encode it in a mathematical construct called a wave function, which enables us to make predictions about information we may gather in future experiments. To leave information, from which we gain all of our knowledge of nature, out of consideration when discussing the nature of the physical world is a gross and outdated oversimplification.

If, however, we accept that information, not matter, lies at the root of quantum mechanics, reductionism, that faithful, flawed handmaiden of science, requires us to ask: what are the fundamental building blocks of information? Zeilinger replies: propositions; Wheeler calls them answers to questions; and the simplest of those, in turn, is an elementary proposition, a yes-or-no question with an answer that's called a 'bit' of information. It is impossible to imagine a simpler question than one that requires a yes-or-no answer.

Here, however, Zeilinger urges caution: we must be clear about the nature of elementary questions. In quantum mechanics, for example, we often ask: 'What is the polarization of this photon?' And the answer may come out: 'Well, there is a 35 per cent probability that it is vertical.' This is certainly not a binary, yes-or-no answer, but the question is not elementary either. A probability, whether it is defined rationally or statistically, is a complex concept. To verify it, many elementary experiments are required, not just one. In the corner of Zeilinger's darkened laboratory, where multicoloured laser beams trace elegant geometric patterns through labyrinths of mirrors, lenses, filters, beam splitters and collimating slits mounted on huge optical tables, there stands a rack of electronic equipment. One of its bays contains a counter that flashes numbers at seemingly random intervals: 27, 28, pause, 29, pause ... *Those* are elementary events. They count whether photons did, or did not, arrive at a certain detector when the slits were set at a certain angle, the mirrors positioned in certain places, the triggers set at certain times. Either a photon arrived, or it did not: Yes or No. It is from these counts that such statements as 'It is vertical 35 per cent of the time' can be computed, and compared with theoretical predictions; but the actual measurements themselves are simple, binary counts. They are expressed as bits.

The idea that all mathematical operations can be expressed in terms of bits has a long history, and underlies the second half of Zeilinger's principle: The smallest amount of information one can give or receive about the world is one bit. The first half of the principle suggests that, since we cannot conceive of information amounting to less than one bit, the simplest physical entity we can understand is described by exactly one bit. Zeilinger calls this irreducible entity an 'elementary system'; but it is not, as one might think, a simple, structureless particle, like an electron, for even the most basic of particles have external attributes, such as energy, direction, position – or intrinsic properties: charge, mass, strangeness – that render them, too, complex. Rather, it is these attributes themselves – such as spin or polarization – that give us elementary systems – not *things*, necessarily, but definitely features of the material world.

The word that connects physical entities with information in Zeilinger's principle is 'carries', and this worries him, because its meaning is not entirely clear. It might be avoided by rephrasing the principle as: 'The information content of an elementary system is one bit,' or, 'An elementary system represents the truth value to one single proposition.' Yet another way to suggest this idea is to define an elementary system as 'a mental construct based on one bit of information' but the exact wording is not really at issue. The point is that one bit of information is associated with an elementary physical system, and vice versa.

The first implication of Zeilinger's principle is that it furnishes an answer to Wheeler's famous question: why the quantum? Why does nature seem granular, discontinuous, quantized into discrete chunks like sand – instead of smooth and continuous like water? The answer is that while we have no idea how the world is really arranged, and shouldn't even ask, we *do* know that knowledge of the world is information; and since information is naturally quantized into bits, the world also appears quantized. If it didn't, we wouldn't be able to understand it. It's both as simple, and as profound, as that.

A second prediction of quantum mechanics that is explained by Zeilinger's principle is the randomness of the outcomes of some measurements – not of all measurements, because certain ones are perfectly predictable. For example, if a hydrogen atom is prepared with its sole electron in its lowest state of energy, then that energy can be measured over and over again with the same result. However, if that same electron is forced to reveal its exact position, the answer will be both unpredictable and, within the broad limits set by the structure of the atom, completely random. It is this essential randomness, so peculiar to quantum mechanics, that distinguishes it from classical physics. The role of chance in the foundations of physics was difficult to accept by older physicists like Einstein, and even Schrödinger resisted it. Zeilinger's principle finally elevates this randomness to the level of a theorem derived from a more fundamental axiom.

The way this comes about is that if the single bit of information in an elementary system is revealed, then there is no more information left over to answer additional experimental questions that are put to the system. Zeilinger gives the example of a particle whose two-valued spin is measured to be 'up', in the vertical z direction. It carries one bit of information – the answer to the question: 'Is the spin up, in the z direction?' If we now ask: 'What is the spin in the x direction?' the answer turns out to be unknown. In fact, the formulas of quantum mechanics predict that the spin in the x direction has a 50:50 probability of being 'up' or 'down', like the random toss of a coin. Zeilinger's principle explains why this must be so. Since the information carried by the spin resides in the z direction, there is no more room for additional information to be collected, so other independent measurements must have random answers.

The most profound consequence of Zeilinger's principle, however, is the illumination it brings to the arcane quantum-mechanical phenomenon of 'entanglement'. Erwin Schrödinger, who coined the German word *Verschränkung* for this effect, and then introduced the English translation himself, went so far as to call it the essence of quantum mechanics. Today, entanglement has been recognized as the key to quantum computing, quantum cryptography, and teleportation. In response to the celebrated critique of quantum mechanics by Einstein, Podolsky and Rosen in 1935, which has reverberated to this very day, Schrödinger defined entanglement as follows: 'Maximal knowledge of a total system does not necessarily include total knowledge of all its parts, not even when these are fully separated from each other.'

Since 1935, many physicists and philosophers of physics have tried to parse this definition, but since the word 'knowledge' is not part of the vocabulary of physics, explanations have invariably reverted to the mathematical terms of quantum mechanics. Zeilinger, by introducing 'information' as the most basic concept of physics, can put the matter much more simply.

Consider the following two-bit scenario: two elementary entities are well separated in space. Each one is described fully by one qubit, and each one carries one bit of information. (Two coins – one showing heads, the other tails – exemplify the situation in classical terms.) Now suppose that the two elementary quantum systems are brought into contact so that they can interact. The same two-bit information content can now be spread out, as it were, between the two of them. If they are subsequently separated again, the information is contained partially in one, partially in the other – even though they may be miles apart. This means that it is impossible, even in principle, to describe the entire system as a mere sum of two subsystems. If the systems are spins, for example, the two propositions that describe them might be: 'The spins measured along the x axis are parallel,' and 'The spins measured along the y axis are opposite.' Once these statements have been made, there is nothing more that can be said – in particular, nothing about the spin of either particle individually, along either one of the two axes. That's what Schrödinger meant when he pointed out that knowledge about the system 'does not necessarily include total knowledge of all its parts'.

The ease with which quantization, randomness and even entanglement are explained speak for the power of Zeilinger's principle. Yet all that is just old wine in new bottles. Physicists tend to sit up and take notice only when a new theory can predict something new; and Zeilinger's principle does.

If quantum communication and quantum computation are to flourish, a new information theory will have to be developed. Its elements will be qubits instead of bits, and in some special circumstances, where it overlaps with the old theory, it will have to give consistent results; but beyond that, a consensus about its formulation has not yet been reached.

One of the most pressing problems of this future theory is

the definition of a quantitative measure of information. Several scientists have pointed out that Shannon's logarithmic formula does not mesh smoothly with the concept of qubits, and on occasion delivers strange results like negative entropy, which would make Boltzmann turn over in his grave. But no one knew precisely what the trouble was, nor how to fix it, until Zeilinger brought his principle to bear on the problem.

In 2001 he and his student Časlav Brukner published a paper entitled 'Conceptual inadequacy of the Shannon information in quantum measurements'. Reviewing Shannon's own justification for his formula, the authors noticed an assumption that appeared so natural at the time that it didn't seem worth mentioning. The assumption is simply that the properties of a system are well defined prior to their observation, and are unaffected by observation. Thus dice, after they have tumbled from their cup, but haven't been observed yet, have definite – albeit unknown – point values. Furthermore, the order in which those values are read off is immaterial: no matter how you determine it, you get the same score. For quantum measurements, however, both these statements are false. The possible results of future observations are not definite beforehand, and further, the order of observations defines and changes their outcomes. Shannon's reasoning in support of his formula therefore fails in the context of quantum mechanics.

Not satisfied with pointing out this problem, Zeilinger and Brukner proceeded to propose a different recipe for measuring information content, which they call 'total information'. It does not contain logarithms, and is actually simpler than Shannon's. Its great virtue is that for one elementary system it yields an information content of one bit, for two elementary systems two bits, and so on to infinity – regardless of how entangled they happen to be.

'Total information' turns out to be the quantum-mechanical implementation of 'information content', abbreviated 'cont', which is found in the collection of classical information measures examined and found acceptable by Kåhre. Philosophers call it 'semantic information content', but it has had many other names since 1912. In the classical (non-quantum) context it is related to the element of surprise. Consider a fair coin. Before you toss it, you know that you are going to be *somewhat* surprised by the outcome. If the coin were two-headed and came up heads, the magnitude of your surprise would be zero. On the other hand, if a miracle happened, and it came up tails instead, your surprise would be complete, which is to say 100 per cent, or one. Accordingly, the 'information content' of a fair coin is set at $\frac{1}{2}$, halfway between zero and one. Since Shannon assigns one bit to a fair coin, the two measures of information differ. Without Shannon's formula, the development of communications and computer technology in the second half of the twentieth century would have been unthinkable. Will 'total information' play a similarly fruitful role in quantum communication and computation in the twenty-first? Time will tell.

Regardless of the answer to this question, Zeilinger's principle is a step in the direction of John Wheeler's prediction: *Tomorrow we will have learned to understand and express all of physics in the language of information.* Wheeler made the prophecy in a famous lecture entitled 'It from Bit', which evolved from a number of presentations to different audiences in 1989. He explained that enigmatic title, which figures among his Really Big Questions when endowed with a question mark, in his customary lapidary style:

> It from bit symbolizes the idea that every item of the physical world has at bottom – at a very deep bottom, in most instances – an immaterial source and explanation; that what we call reality arises in the last analysis from the posing of yes-no questions and the registering of equipment-evoked responses; in short, that all things physical are information-theoretic in origin and this in a *participatory universe*.

Zeilinger's accomplishment can be seen in this light as translating Wheeler's words into the mathematical terms of theoretical physics, and the first steps towards deriving rigorous consequences from them. He was helped in this endeavour by the recent addition of the word 'qubit' to the arsenal of quantum mechanics: the qubit is the quantum-mechanical device physicists have developed in order to describe Wheeler's 'item of physical reality', and Zeilinger's 'elementary system'. The qubit is much richer in content than the bit. The power of Zeilinger's principle derives precisely from its confrontation of the qubit – the irreducible building block of inanimate matter – with the bit – the fundamental quantum of human knowledge. That they should find themselves in one-to-one correspondence is both the simplest assumption one can possibly make, and a deep insight.

Zeilinger, being at heart an experimentalist, puts his faith in the bit, the nugget of certainty that nature allows us to extract from a qubit. The bit corresponds to a click in his counter or a dot in his photograph. It represents nature as he perceives and measures it. Being a theorist, I look with awe at the other side of the equation, the qubit, from which the bit was squeezed. The qubit, which floats through my mind in the form of a soft, translucent sphere, a peeled, seedless grape shimmering indistinctly in all the colours of the rainbow at once, is an inexhaustible source of possibilities from which only one can finally be realized. It is ripe with infinite surprise, for the bit is contained in it in an irreducibly random, unpredictable way. Grabbing the qubit, measuring it, and reducing it to a definite colour, or a comprehensible Yes or No, or a numerical zero or one, is an original act of creation on my part, undetermined by the past, irreproducible in the future. To me, the qubit is the ultimate source of wonder.

Notes

Prologue

pp. xi–xii '"Every it – every particle ..."' John A. Wheeler, *At Home in the Universe*, Springer, 1992, p. 296.

p. xii '"... in the language of information".' Ibid., p. 298.

p. xiii '... (a fine point in the theory ...) ...' H. C. von Baeyer and J. Callaway, 'The effect of point imperfections on Lattice Thermal Conductivity', *Physical Review* 120, 1149 (1960).

p. xiii '"... conflict in their views."' Joseph Callaway, 'Mach's Principle and Unified Field Theory', *Physical Review* 96, 778 (1954).

p. xiv '... IT FROM QUBIT?' Gerard J. Milburn, *The Feynman Processor*, Perseus Books, 1998, p. 37.

Chapter 1: Electric Rain

p. 4 '... more information in the next three years ...' *New York Times* (hereafter referred to as *NYT*), 15 January 2001, p. C1.

p. 5 'When IBM discovered ...' Marshall McLuhan, *Understanding Media*, McGraw-Hill, 1964, p. 9.

p. 5 '... will not be reached until the 2020s ...' Neil Gershenfeld, *The Physics of Information Technology*, Cambridge University Press, 2000, p. 161.

p. 6 '... an ocean of data ...' George Johnson, *NYT*, 2 February 2001, p. 4.

p. 7 '... who lack access to clean drinking water.' Samuel R. Berger, *NYT*, 20 January 2001, p. A19.

p. 7 '... will become prohibitive.' *Science News* 16, p. 309 (2002).

p. 7 ' "Pulling Diamonds from the Clay" ' In Peter J. Denning, ed. *Talking Back to the Machine: Computers and Human Aspiration*, Copernicus Springer-Verlag, 1999.

p. 8 'Even the brain ...' Ian Glynn, *An Anatomy of Thought*, Oxford University Press, 1999, p. 106.

p. 9 '... its computer-age replacement ...' ASCII (American Standard Code for Information Interchange) assigns specific strings of zeroes and ones to hundreds of alphanumeric and typographical symbols in the principal languages of the world.

p. 9 '... modern methods of communication.' Barbara King, *The Information Continuum*, University of Washington Press, 1994.

p. 10 '... Feynman's monumental textbook ...' However, information does figure prominently in the posthumous *Feynman Lectures on Computation*, Tony Hey and Robin W. Allen, eds., Westview, 1996.

Chapter 2: The Spell of Democritus

p. 13 '... demonstrated with a swift kick.' James Boswell, *The Life of Samuel Johnson*, Encyclopaedia Britannica Great Books, vol. 44, 1952, p. 134.

p. 13 'Only a few fragments of his work ...' Bernard Pullman, *The Atom in the History of Human Thought*, Oxford University Press, 1998, p. 31.

pp. 13–14 ' "Sweet is by convention ..." ' Erwin Schrödinger, *Nature and the Greeks & Science and Humanism*, Cambridge University Press, 1996, p. 89.

p. 15 'The fragment ... reads in full: ...' Ibid.

p. 15 'No theory of physics ...' Quoted in Eugene Hecht, *Perspectives in Physics*, Addison-Wesley, 1980, p. vi.

p. 16 'If, in some cataclysm ...' Richard Feynman, *Lectures on Physics*, Addison-Wesley, 1963, vol. I, p. 1–2.

Chapter 3: In-Formation

p. 18 ' "No problem," quipped his brother ...' Quoted in Henning Genz, *Wie die Zeit in die Welt kam*, Carl Hanser Verlag, Munich, 1996, p. 274.

pp. 21–2 'We have learned ... Geometry.' D'Arcy Wentworth Thompson, *On Growth and Form* (abridged), Cambridge University Press, 1961, p. 269.

p. 22 '... in the broadest possible way.' Paul Young, *The Nature of Information*, Praeger, 1987, p. 52.

p. 22 '... these exquisite forms.' Andrew Hodges, *Alan Turing – The Enigma of Intelligence*, Unwin Paperbacks, 1983, p. 492.

p. 23 ' "Geometrizing a theory" ...' David Ruelle, 'Conversations on Mathematics with a Visitor from Outer Space', in *Mathematics: Frontiers and Perspectives 2000*, International Mathematical Union, 2000, p. 251.

p. 24 'The aim of science ...' Henri Poincaré, *Science and Hypothesis*, Dover, 1952, p. xxiv.

p. 24 ' "We cannot think of any object ..." ' Ludwig Wittgenstein, *Tractatus*, 2.0121. Quoted by David Mermin in 'What is quantum mechanics trying to tell us?' *American Journal of Physics*, vol. 66, p. 753 (1998).

p. 24 ' "... intricate relationships seeking a form." ' Italo Calvino, *Invisible Cities*, Harcourt Brace Jovanovich, 1972, p. 76.

p. 25 'Cicero used the verb "inform" ...' Oxford Latin Dictionary, 1968.

p. 25 'It possesses "informative power" ...' Robert Wright, *Three Scientists and their Gods*, Times Books, 1988, p. 95.

Chapter 4: Counting Bits

p. 28 '... eight possible strings of heads and tails.' If '0' means tails and '1' heads, three coins can come up 000, 001, 010, 011, 100, 101, 110, or 111. Conveniently, these happen to be the numbers 0 to 7 in binary code. In this notation, the different outcomes of coin tosses 'count themselves'.

p. 29 'With good strategy ... twenty factors of two.' N. J. A. Sloane and A. D. Wyer, eds., *Claude Elwood Shannon: Collected Papers*, IEEE Press, 1993, p. 215.

p. 30 ' "The beginning of modern science ..." ' François Jacob, *The Possible and the Actual*, Pantheon, 1982, p. 10.

p. 32 ' "If we are smart ..." ' Wright, op. cit., p. 95.

p. 33 '... an IGUS ...' M. Gell-Mann, *The Quark and the Jaguar*, Little, Brown, 1994, p. 155.

p. 33 ' "... the real role of the observer." ' Tom Siegfried, *The Bit and the Pendulum*, Wiley, 2000, p. 175.

Chapter 5: Abstraction

p. 36 ' "In its general structure ..." ' Ernst Cassirer, *The Philosophy of Symbolic Forms*, Yale University Press, 1957, p. 460.

p. 37 '... "flight from common sense" ...' Laurie Brown et al., eds., *Twentieth Century Physics*, AIP Press, 1995, p. 2018.

p. 39 ' "*The essential reality* ..." ' Quoted in Heinz Pagels, *The Cosmic Code*, Simon and Schuster, 1982, p. 269.

p. 40 'He now hopes ... turn out to be.' Steven Weinberg, *Facing Up*, Harvard University Press, 2001, p. 97.

p. 40 '... a hundred thousand times more precise.' Graham Farmelo, ed., *It Must Be Beautiful*, Granta, 2002, p. 145.

Chapter 6: The Book of Life

p. 44 '... Planck/Einstein hypothesis ...' Helge Kragh, *Quantum Generations*, Princeton University Press, 1999, p. 65.

p. 46 ' "The Lord God is subtle ..." ' Alice Calaprice, *The Expanded Quotable Einstein*, Princeton University Press, 2000, p. 241.

p. 47 'The three traits ...' Peter Raven and George Johnson, *Biology*, Times Mirror/Mosby College Publishing, 1989, p. 247.

p. 47 ' "It doesn't make the slightest difference ..." ' Burton Feldman, *The Nobel Prize*, Arcade Publishing, 2000, p. 253.

p. 47 ' "Must we geneticists ..." ' Ibid.

p. 47 '... the book *What is Life?* ...' Erwin Schrödinger, *What is Life? – The Physical Aspect of the Living Cell*, Macmillan, 1945.

p. 51 '[The] molecular model ...' Ibid., p. 68.

p. 52 '... the entire length of a chromosome.' Raven and Johnson, op. cit., p. 204.

p. 52 'In *What is Life?* ...' Schrödinger, op. cit., p. 28.

Chapter 7: A Battle Among Giants

p. 54 ' "... is 'the lawn' what we see ..." ' Italo Calvino, *Mr Palomar*, Harcourt Brace Jovanovich, 1985, p. 32.

p. 56 '... and the battle was joined.' Weinberg, op. cit., p. 13.

p. 56 '... called *petit* and *grand mal*.' Ibid., p. 111.

p. 57 '... Anderson's book review ...' Philip W. Anderson, review of Weinberg, op. cit., *Physics Today*, July 2002, p. 56.

p. 58 '... "the primary and essential activity of science" ...' Edward O. Wilson, *Consilience*, Knopf, 1998, p. 54.

p. 59 'Economists ... measure information in dollars.' See, for example, the discussion of the value of information in Robert S. Pindyck and Daniel L. Rubinfeld, *Microeconomics*, 5th edn, 2001, p. 164.

Chapter 8: The Oracle of Copenhagen

p. 63 '... Bohr was forced ... to reject his own model.' Abraham Pais, *Niels Bohr's Times*, Clarendon Press, 1991, p. 202. My essay about this book in the *Gettysburg Review*, Summer 1990, contains some of the material in this chapter.

p. 64 ' "... not sure there is no real problem." ' Quoted in David Park, *The How and the Why*, Princeton University Press, 1988, p. 341.

p. 65 ' "Our task is ..." ' Pais, op. cit., p. 446.

p. 65 ' "There is no quantum world ..." ' Ibid., p. 426.

Chapter 9: Figuring the Odds

p. 69 '... a magazine article that described it.'
<http://www.wiskit.com/marilyn.gameshow.html> and many other sites.

p. 72 '... they gave up and dropped the subject.' Paul Hoffman, *The Man Who Loved Only Numbers*, Hyperion, 1998, p. 249.

p. 72 '... the other child is also a boy?' Martin Gardner, *The 2nd Scientific American Book of Mathematical Puzzles and Diversion*, Simon and Schuster, 1961, p. 226, points out that the answer depends on how the father is chosen. I assume that he is picked at random from a collection of families with two children of whom at least one is a boy.

p. 73 '... vast amounts of missing information.' Claude E. Shannon and Warren Weaver, *The Mathematical Theory of Communication*, University of Illinois Press, 1949, p. 95.

p. 75 ' "... the probability of any particular proposition." ' Quoted in Ralph Baierlein, *Atoms and Information Theory*, W. H. Freeman, 1971, p. 11. Baierlein's book is my principal reference for the remainder of this chapter.

p. 76 ' "... those for predicting the weather".' Ibid., p. 60.

p. 78 '... tells the amazing story.'

	Cancer test positive	Cancer test negative	Total
Cancer patients	99	1	100
Healthy people	99	9801	9900
Total	198	9802	10,000

Bayes's theorem: $(1/100) \times [(99/100)/(198/10,000)] = \frac{1}{2}$

p. 79 ' "It's a real-life example," ...' David Leonhardt, *NYT*, 28 April 2001, p. A17.

p. 80 '... the difficulty dissolves.' Carlton M. Caves, 'Resource Material for Promoting the Bayesian View of Everything', <http://info.phys.unm.edu/~caves/thoughts2.2.pdf>

Chapter 10: Counting Digits

p. 81 '... the computational power of astronomers ...' Carl B. Boyer, *A History of Mathematics*, Princeton, 1985, p. 346.

p. 83 '... counts the digits ... at least approximately.' Sometimes, as in the case of 1000, this recipe is off by one digit, but that hardly matters when you are dealing with the Gargantuan and Lilliputian numbers of science, and increasingly of modern life in general.

p. 86 '... another more recent book ...' Fred Adams and Greg Laughlin, *The Five Ages of the Universe*, The Free Press, 1999, p. 237.

p. 87 ' "... it passes in a twinkling to a pensioner." ' Sheldon Glashow, *From Alchemy to Quarks*, Brooks/Cole, 1993, p. 26.

p. 87 '... long-standing puzzles of modern physics.' Frank Wilczek, 'Scaling Mount Planck', *Physics Today*, June 2001, p. 12; November 2001, p. 12; August 2002, p. 10.

Chapter 11: The Message on the Tombstone

p. 92 '... as ... Sir Arthur Eddington put it.' Arthur Eddington, *The Nature of the Physical World*, Macmillan, 1929, p. 74.

p. 96 'When Max Planck ...' According to W. T. Grandy, Jr, in 'Resource Letter on Information Theory in Physics', *American*

Journal of Physics, vol. 65, p. 467 (1997), Planck meant W to stand for probability (*Wahrscheinlichkeit* in German). However, in modern usage probability is a number between zero and one, and has a negative log. This would render the entropy negative, contrary to common usage. Reading W as *number of ways* makes everything come out fine.

p. 97 '... immoderate enthusiasm in others.' David Ruelle, *Chance and Chaos*, Princeton University Press, 1991, p. 101.

p. 98 '... our lack of information.' Shannon and Weaver, op. cit., p. 95.

Chapter 12: Randomness

p. 101 '... real number written in binary code ...' Binary numbers work just like decimal numbers, except that their digits run from 0 to 1 instead of from 0 to 9. Thus the counting numbers 0, 1, 2, 3, 4, 5, 6, 7, 8, 9, 10, 11 become, in binary code, 0, 1, 10, 11, 100, 101, 110, 111, 1000, 1001, 1010, 1011.

p. 101 '... David Champernowne ... found one.' Edward Beltrami, *What Is Random?*, Springer/Copernicus, 1999, p. 33.

p. 101 '... then a raw beginner.' Hodges, op. cit., p. 388.

p. 101 '... at age eighty-eight.' www.nuff.ox.ac.uk/Users/ SHEPHARD/OXECONGROUP/Champernowne.html

p. 102 '... several mathematicians ...' Dana Mackenzie, 'On a Roll,' *New Scientist*, 6 November 1999, p. 44.

p. 102 '... "algorithmic complexity" ...' Another word for it is 'Kolmogorov complexity'. There are many definitions of complexity, of which this one seems to be the most fruitful.

p. 105 '... not very good at generating truer andomness.' Beltrami, op. cit., p. 125.

p. 105 '... which beat his colleague's.' Sloane and Wyer, eds., op. cit., p. xvii.

p. 105 '... human frailty betrayed them.' Simon Singh, *The Code Book*, Doubleday, 1999, p. 164.

p. 105 '... "in a state of sin," ...' Colin P. Williams and Scott H. Clearwater, *Ultimate Zero and One*, Copernicus, 2000, p. 135.

p. 105 '... "Random Numbers Fall Mainly in the Plane" ...' Beltrami, op. cit., p. 81.

p. 108 '... were able to solve it.' Mackenzie, op. cit., p. 44.

p. 109 '... back to its previous value.' W. H. Żurek, 'Algorithmic randomness and physical entropy', *Physical Review*, A 40, p. 4731 (1989).

Chapter 13: Electric Information

p. 111 '... annihilation of space had been accomplished.' Oliver W. Larkin, *Samuel F. B. Morse and American Democratic Art*, Little, Brown, 1954, p. 151.

p. 112 ' " ... *one neighbourhood* of the whole country." ' Paul J. Staiti, *Samuel F. B. Morse*, Cambridge University Press, 1989, p. 223.

p. 112 ' " ... as far as our planet is concerned." ' McLuhan, op. cit., p. 3.

p. 113 '... force of the prevailing populism.' Staiti, op. cit., p. 235.

p. 114 '... an insurmountable bottleneck.' John R. Pierce, *An Introduction to Information Theory*, Dover 1980, p. 24.

p. 115 '... style called "telegraphese" ...' In their heyday, telegrams were drastically condensed by means of dictionaries such as the *Rudolf Mosse Code* (1922), which listed fictitious five-letter words that stood for entire phrases useful in commercial communications.

p. 115 '... drew four symbols each.' Pierce, op. cit., p. 25.

p. 116 '... founding father of the cyber age.' *NYT* Obituary, 27 February 2001.

p. 117 '... "totally useless things." ' Sloane and Wyer, eds., op. cit., p. xvi.

p. 120 '... by no more than 15 per cent.' Pierce, op. cit., p. 25.

Chapter 14: Noise

p. 124 '... or log S/N expressed in bits.' In order to prevent this quantity from becoming negative when S/N happens to be less than one, Shannon actually replaced it by log (1 + S/N).

p. 125 'If one pulse can carry log S/N bits ...' A more precise analysis substitutes the bandwidth (the range of possible frequencies) for the highest attainable frequency. My crude way is correct only if the lowest transmitted frequency happens to be zero.

p. 126 '... from biology to quantum physics.' Frank Moss and Kurt Wiesenfeld, 'The Benefits of Background Noise', *Scientific American*, August 1995, p. 66.

Chapter 15: Ultimate Speed

p. 131 '... *backwards in time.*' John Stachel and David C. Cassidy, *The Collected Papers of Albert Einstein*, vol. 2, Princeton University Press, 1989, p. 424.

p. 131 '... play havoc with cause and effect.' In particular, an iris wave sent out in the backward direction by a rocket moving forward at half the speed of light would appear to an observer on the ground to be moving backwards in space and *backwards in time* at three and a half times the speed of light. Cf. Claude Kacser, *Introduction to the Special Theory of Relativity*, Prentice-Hall, 1967, p. 81.

p. 131 '... Arnold Sommerfeld ... gave a short lecture ...' Leon Brillouin, *Wave Propagation and Group Velocity*, Academic Press, 1960, p. 12.

p. 133 '... even before it entered it.' A. Dogariu, A. Kuzmich, and L. J. Wang, 'Transparent anomalous dispersion and superluminal light-pulse propagation at a negative group velocity', *Physical Review* A 63, 053806–1, 2001.

p. 134 '... that limits the speed of information to c.' A. Kuzmich, A. Dogariu, L. J. Wang, P. W. Milonni, and R. Y. Chiao, 'Signal Velocity, Causality, and Quantum Noise in Superluminal Light Pulse Propagation', *Physical Review* Letters 86, p. 3925, 2001.

Chapter 16: Unpacking Information

p. 136 '... are developed and verified.' Reproduced and discussed by Gerald Holton, *The Advancement of Science, and its Burdens*, Cambridge University Press, 1986, p. 31.

p. 138 ' "The grand aim of all science ... " ' Quoted in Calaprice, op. cit., p. 256.

p. 138 ' "... breathing in and breathing out." ' Quoted in Wilson, op. cit., p. 36.

p. 139 '... the *Einstein equation.*' For what it's worth, here it is: $R_{\mu\nu} - Rg_{\mu\nu}/2 + \lambda g_{\mu\nu} = -8\pi GT_{\mu\nu}$. As a graduate student I spent countless nights wrestling with this magnificent monster.

p. 143 'In search of escapes ...' A brief summary is given in Siegfried, op. cit., p. 175.

Chapter 17: Bioinformatics

p. 145 '... than go to make up an earthworm genome.' The consensus estimate became 30,000, considerably below the original guess of 100,000. Recently a number around 60,000 has been proposed. Clearly the genome is a work in progress!

p. 148 '... a remarkable similarity shows up.'
http://www.uni-mainz.de/%7Ecfrosch/bc4s/example.html

p. 150 'One promising avenue ...' Pierre Baldi and Søren Brunak, *Bioinformatics: the machine learning approach*, 2nd edn, MIT, 2001.

p. 151 'The information needed ...' This statement can be quantified. Cf. D. T. Gregory, 'Algorithmic complexity of a protein,' *Physical Review* 54, R39 (1996).

Chapter 18: Information is Physical

p. 153 'In order to consider ...' H. C. von Baeyer, *Warmth Disperses and Time Passes – The History of Heat*, Modern Library, 1999, p. 39. This is my source for Carnot and his work.

p. 157 '... skittering about in a warm room.' Combining Clausius's definition of entropy as heat over temperature with the formula on Boltzmann's grave and Shannon's definition of a bit, that amount comes out to be kTlog2 units of energy. Here k is Boltzmann's constant, T the temperature of the computer, and 2 the number of messages a bit can convey.

p. 157 '"... puzzles in the sociology of science".' Rolf Landauer, 'Zig-Zag Path to Understanding', *Proceedings of the Workshop on Physics and Computation*, Dallas, TX, 17–20 November 1994, IEEE Computer Society.

p. 158 '... a famous article ...' Rolf Landauer, 'Information is Physical', *Physics Today*, May 1991, p. 23.

p. 159 '... without a single erasure.' In our example, 5 and 3 are input data, and are stored in a notebook outside the computer. Imagine the computer as some kind of frictionless abacus. It stores the 5 and the 3, and then performs the internal operation of addition. After it arrives at the answer, 8, that number is copied to the notebook. Finally the computer is instructed to subtract 5, 3 and 8 from the relevant internal memory cells.

The slate is thus wiped clean by frictionless computation, not by erasure.

p. 159 'Computation is cool!' Milburn, op. cit. p. 123.

Chapter 19: The Quantum Gadget

p. 166 'The experiment in question ...' For some of this material I am indebted to my student, Rachele Dominguez, *Using Information to Reformulate Quantum Mechanics*, Senior Honors Thesis, College of William and Mary, 2002.

Chapter 20: A Game of Beads

p. 178 '... a term coined by Schrödinger ...' Erwin Schrödinger, 'Discussion of probability relations between separated systems', *Proceedings of the Cambridge Philosophical Society* 31, 555 (1935).

p. 179 '... rises dramatically.' For N beads, that probability is given by $[\cos(90°/N)]^2$. For $N = 6$, the angle is 15 degrees, and the square of the cosine 0.9330. As N increases, the angle tends toward zero, and the value of the cosine rapidly approaches unity.

Chapter 21: The Qubit

p. 183 '... quantum version of the bit was invented.' Siegfried, op. cit., p. 34.

p. 184 '... quantum information without the qubit.' The Cornell physicist David Mermin objects to the orthography of qubit on the grounds that in English 'qu' should be followed by a vowel. He urges, without much hope for success, adoption of the alternative spelling 'qbit'.

p. 190 '... and 0 at the south pole.' Williams and Clearwater, op. cit., p. 14.

Chapter 22: Quantum Computing

p. 193 '... a delicate, fey neurasthenic ...' Neurasthenic: a person afflicted with a type of neurosis characterized by a wide variety of symptoms including worry and localized pains without apparent objective causes.

p. 193 ' "... the ultimate nature of physical law." ' Quoted in Milburn, op. cit., p. 192.

p. 194 '... 2020s has been suggested ...' Williams and Clearwater, op. cit., p. 6.

p. 197 '... that guarantee unbreakable communications.' Ibid., p. 235.

p. 202 '... or any other quantum-mechanical system.' Ibid., p. 55.

p. 203 '... a scale a million times smaller.' S. Somaroo et al., 'Quantum simulations on a quantum computer', *Physical Review* Letters 82, 5381 (1999).

Chapter 23: Black Holes

p. 204 ' "We will never understand ..." ' Quoted by Edward Teller in *Niels Bohr – A Centennial Volume*, A. P. French and P. J. Kennedy, eds., Harvard University Press, 1985, p. 181.

p. 205 '... a "perfect crime".' Jacob D. Bekenstein, 'The Limits of Information', *Studies in the History and Philosophy of Modern Physics* 32, p. 511 (2002).

p. 210 '... we run the danger of trespassing ...' Stephen W. Hawking and Werner Israel, 'Black Holes', in *Encyclopedia of Physics*, 2nd edn, p. 104 (1991).

p. 211 '... "one of the major questions" ...' Quoted in Brian Greene, *The Elegant Universe*, Norton, 1999, p. 343.

Chapter 24: Bits, Bucks, Hits and Nuts

p. 215 '... the Pythagoras of information science.' Quoted in Jan Kåhre, *The Mathematical Theory of Information*, Kluwer Academic Publishers, 2002, p. 8, and http://www.matheory.info

p. 216 ' "... As a mathematical quantity ..." ' Ibid., p. 301.

p. 216 '... and quantum mechanics.' W. K. Wooters, *The Acquisition of Information from Quantum Measurements*, PhD dissertation, University of Texas at Austin, 1980.

p. 216 '... has not found many supporters.' B. R. Frieden, *Physics From Fisher Information – a Unification*, Cambridge University Press, 1998. For a critique see D. A. Lavis and R. F. Streater, 'Essay Review of Physics From Fisher Information', *Studies in History and Philosophy of Modern Physics*, vol. 33, p. 327, June 2002.

p. 217 '... biased towards Shannon's results.' Kåhre, op. cit., p. 107.

Chapter 25: Zeilinger's Principle

p. 223 ' "... Forever Degenerating Influence on Others." ' Engelbert Broda, *Ludwig Boltzmann*, Ox Bow Press, 1983, pp. 98 and 141.

p. 223 ' "... Mach's philosophical writings." ' Quoted from Einstein's 'Autobiographical Notes' by Silvio Bergia in *Einstein, A Centenary Volume*, A. P. French, ed., Heinemann, London, 1979, p. 84.

p. 223 ' "Physics does not consist ..." ' Erwin Schrödinger, cited by Walter Moore, *Schrödinger: Life and Thought*, Cambridge University Press, 1989, p. 226.

p. 224 ' "... also the questions of human life." ' Marcus E. Fierz in *The Life and Times of Modern Physics: History of Physics II*, American Institute of Physics, 1992, p. 298.

p. 226 'He reported his findings ...' <http://www.ap.univie.ac.at/users/Anton.Zeilinger/philosop.html>, retrieved 6/16/00.

p. 227 ' "... the guys resist admitting this ..." ' Ibid., p. 15.

p. 227 'On 5 January he submitted a paper ...' *Foundation of Physics*, vol. 29, No. 4, p. 631 (1999).

p. 228 'The idea that all ...' It is attributed to Leibniz (1697) by Tino Gramß et al., *Non-Standard Computation*, Wiley-VCH, 1998, p. 1.

p. 230 ' "Maximal knowledge ..." ' Erwin Schrödinger cited in J. A. Wheeler and W. H. Żurek, *Quantum Theory and Measurement*, Princeton University Press, 1983, p. 152.

p. 232 ' "Conceptual inadequacy ... in quantum measurements".' *Physical Review* A 63, 022113 p. 1, 2001.

p. 232 '... simpler than Shannon's.' For a coin with a probability of coming up heads p, the formula is simply $I = (2p - 1)^2$ bits. This implies that an honest coin, with $p = \frac{1}{2}$, makes no prediction about the future. Once it is thrown, however, and p becomes zero or one, it conveys one bit of information.

p. 232 '... found acceptable by Kåhre.' Kåhre, op. cit., p. 88.

p. 232 '... since 1912 ...' Ibid., p. 501. Thanks to Jan Hajek for this reference.

p. 234 'It is ripe with infinite surprise ...' This point is made most eloquently in Milburn, op. cit.

Index